# Solutions Manual

*to accompany*

# Research Design
# and
# Statistical Analysis

## Second Edition

# Solutions Manual

*to accompany*

# Research Design
# and
# Statistical Analysis

## Second Edition

Jerome L. Myers
and
Arnold D. Well
*University of Massachusetts*

**Ψ Psychology Press**
Taylor & Francis Group

New York   London

First Published by
Lawrence Erlbaum Associates, Inc., Publishers
10 Industrial Avenue
Mahwah, New Jersey 07430

Transferred to Digital Printing 2009 by Psychology Press
270 Madison Ave, New York NY 10016
27 Church Road, Hove, East Sussex, BN3 2FA

ISBN: 0-8058-4438-4

**Publisher's Note**
The publisher has gone to great lengths to ensure the quality of this reprint
but points out that some imperfections in the original may be apparent.

# Contents

# Solutions Manual

*to accompany*

# Research Design
# and
# Statistical Analysis

## Second Edition

# Chapter 1

**1.1**

(a) The study is an experiment. Participants were randomly assigned to each condition.

(b) The sample might best be characterized as a random sample of (presumably) overweight volunteers. The volunteer aspect is important because it is not clear that the effects of the diets would be the same if used by individuals less motivated to lose weight.

(c) One alternative approach is to observe individuals who have selected a particular diet. This might be done by advertising for individuals who are on the diets of interest, requesting they volunteer as participants in a study. This not as attractive an option as the experiment for several reasons. For example, baseline information on weight and health prior to the start of the diet is likely to be lacking. One or more of the targeted diets may not be well represented in the volunteer sample. Individuals who chose different diets may differ systematically with respect to other factors that may influence weight loss.

**1.2**

(a) It is reasonable to assume that the results hold for children from a similar home and school environment, though before drawing conclusions about the two methods, it would be wise to test generality by replicating the experiment at other similar schools. With respect to the broader population of third graders, we should be still more cautious about generalizing until further research is carried out. Factors such as experience with computers and parental involvement in arithmetic learning may be different for other segments of the population.

(b) A major problem in such studies is equating the amount of practice in the two conditions. Another variable is teacher experience and competence. Assigning the same teacher to both classes is a possible solution but not necessarily the best one if the teacher is biased in favor of either method. The bottom line is that considerable effort must be taken both with the materials and the teachers prior to the experiment.

**1.3**

(a) There are several factors other than long hours in day care that may increase aggressive behavior. Two possibilities are that (1) parents may be more likely to place more aggressive children in day care in order to reduce unpleasant interactions, or to provide a socializing experience; (2) parents who are more stressed may foster aggression in their children and may place those children in day care for longer hours in order to reduce their stress.

(b) Ideally, a measure of aggressive behavior prior to the first enrollment in day care should be obtained. If variation in aggressive behavior as a function of hours/day in day care manifests itself only after the first placement in day care, there is evidence that time in day care has an effect. (Of course this doesn't reveal whether the relation is due to the day care environment, or absence from parents.) If children who were more aggressive after placement in day care were also more aggressive prior to first placement, there is an indication that factors such as those listed in part (a) are involved. Other measures such as measures of parental stress and rated quality of the home environment may also prove useful. Using the regression procedures referred to in this chapter, and developed in Chapters19-21, we can investigate whether such measures predict variability in aggressive behavior beyond that predicted by hours in day care.

(c) An experiment could be performed by randomly assigning children to day care or to home care. To further reduce the influence of nuisance variables, care might be taken to equate the groups with respect to such factors as parents' working hours, parents' rating of their stress, and whether the child has one or two parents in the home. The advantage of the experimental approach is that random assignment ensures that there will be no systematic bias due to uncontrolled factors and several factors may be controlled by equating the groups with respect to them. The disadvantage is that the approach is both unethical and impractical. It is unethical because if one treatment does cause problems, the researcher has subjected half of the children to an experience that may have a negative effect on the child's future. The experiment is impractical because it is doubtful that parents will allow a researcher to decide the preschool experience of their children.

**1.4** As any representative of a tobacco company might argue, there are many factors that might lead to both cigarette smoking and lung cancer. For example, individuals in jobs that expose them to carcinogens may tend to smoke more. Individuals who are emotionally less stable or more depressed may smoke more and may be more susceptible to cancer because of psychological effects on their physical health. Although a true experiment would be unethical, a "natural experiment" is possible, and in fact at least one has been conducted. An investigation of the incidence of lung cancer in Amish and non-Amish residents of Lancaster county revealed that the Amish, who are nonsmokers, had a *zero* incidence of lung cancer. Although it is conceivable that other factors in their life style might have caused this, the result is an impressive contribution to the case against smoking. Other evidence comes from many studies of lung tissue from individuals who died of lung cancer; the evidence is correlational but consistent: there is a strong correlation between damage to cellular structure and the amount of cigarette smoking. Consistent with this evidence is the finding that various chemical components of cigarettes can be shown to have effects on cellular structure, leading to deviations from the normal state of those cells.

**1.5** Less education may cause lower income, less interesting jobs, and poorer health (because of lower income or because of less knowledge about practices that promote health). All of these consequences may, in turn, increase depression. Thus there is some credibility to the idea that the difference in depression scores between high-school educated and college educated individuals is affected by education. However, further examination suggests other possibilities. For example, if the high-school educated group came from families with less income, they may have been more likely to receive less education and also to be more depressed. It is also possible that other factors in the family environment mediated differences in both education and depression scores. Other factors may also be important. For example, from the data available on the CD, we might try to determine whether there are systematic age differences between the two groups and, if so, whether depression is related to age within each educational level. If so, the difference in depression scores might be the result of differences in age.

### 1.6

(a) The independent variable is whether individuals were smokers or nonsmokers. The dependent variable is frequency of lung cancer.

(b) The independent variable is discrete with two values, smoke or don't smoke. The dependent variable is also discrete with possible integer values ranging from zero to $n$, the number of individuals in the group.

(c) There are many additional measures that would aid in interpreting the results of such a study. Here are a few suggestions. We would want to know how the number of years (continuous) each participant in the smoking group had been smoking at the onset of the study. We would want a record of the average daily number of cigarettes smoked (continuous). The incidence of lung cancer (discrete) and the time of onset of cancer (continuous) could be correlated with these variables. Presumably, the incidence of cancer would be highest for the subset of individuals who had been smoking longest and for those individuals who smoked more cigarettes each day. It also would be useful to have age (continuous) and gender (discrete) information for each individual. in order to investigate whether – all other things being equal – the effects of cigarette smoking are related to these variables. In addition, information about other aspects of the participant's health should be obtained at the start of the study and probably recorded at regular intervals. For example, blood pressure levels (continuous) might be recorded. Measures of health and fitness may be related to smoking factors.

## Chapter 2

**2.1** (a) $\overline{Y} = 30.562$; (b) $\tilde{Y}$ (median) $= 33.5$; (c) $(\Sigma Y)^2 = 239{,}121$; (d) $\Sigma Y^2 = 16{,}311$; (e) $9.543$; (f) $H_L = 23.5$, $H_U = 37.5$.

### 2.2

(a) First convert the $Y$ scores to $z$ scores by $z = (Y - \overline{Y})/s$; this transforms the data to a scale with $\overline{Y} = 0$ and $s = 1$. Then, multiply by 15 and add 100.

(b) Call the new score $X$. Then $\tilde{X} = 104.617$, $H_L = 88.889$, and $H_U = 110.905$.

**2.3** Outliers in box or stem-and-leaf plots and the shape of a normal probability plot suggest a heavy-tailed distribution in data set (a). Both a stem-and-leaf (or a histogram) and a normal probability plot indicate that data set (c) is skewed to the right. Data set (b) appears to be normally distributed.

**2.4** The depth of the median is 10.5; therefore, its value lies between 16 and 17, or 16.5. Applying formulas for depth of other positions,

$$d_{fourth} = \frac{[d_M]+1}{2} = 11/2 = 5.5$$

The fifth and sixth smallest scores are both 14 and the fifth and sixth largest scores are 32 and 31. Therefore, F = 14 and 31.5. Similarly,

$$d_{eighth} = \frac{[d_{fourth}]+1}{2} = 6/2 = 3$$

and E = 10 and 35. It should be evident that the median lies closer to the lower fourth and eighth, suggesting a long tail to the right.

**2.5**

(a) If the mean of six scores is 47, the sum must be 6 x 47, or 282. However, $\sum_{i=1}^{5} Y_i = 225$. Therefore, the 6$^{th}$ score must be 282 - 225, or 57.

(b) The mean of the original 5 scores is 45. Adding a score equal to the mean will yield the smallest variance because the variance is the sum of squared deviations about the mean.

**2.6**

(a) $\sum_{i=1}^{4}(X_i + Y_i) = \sum_{i=1}^{4} X_i + \sum_{i=1}^{4} Y_i = 19 + 53 = 72$.

(b) $\sum_{i=1}^{5} X_i^2 = 6^2 + 5^2 + ... + 11^2 = 232$.

(c) $\left(\sum_{i=1}^{5} X_i\right)^2 = 30^2 = 900$.

(d) $\sum_{i=1}^{5} X_i Y_i = (6)(7) + (5)(11) + ... + (11)(9) = 315$.

(e) $\sum_{i=1}^{5}(X_i + aY_i^2 + ab) = \sum_{i=1}^{5}(X_i) + a\sum_{i=1}^{5}(Y_i^2) + 5ab$
$$= 30 + (5)(888) + (5)(3)(9) = 4,575.$$

**2.7**

(a) $\overline{Y}._1 = (7+31+...+35)/5 = 22$ .

(b) $\overline{Y}_2. = (31+15+12)/3 = 19.333$ .

(c) $\overline{Y}.. = (7+31+...+19+4)/15 = 21.067$ .

(d) $\sum\limits_{i=1}^{5}\sum\limits_{j=1}^{3} Y_{ij}^2 = 7^2 + 31^2 + ... + 19^2 + 4^2 = 9,422$ .

(e) $\sum\limits_{j=1}^{3} \overline{Y}._j^2 = 22^2 + 30.6^2 + 10.6^2 = 1,532.72$ .

**2.8** Consider location first; whether we look at the mean or the median, the $X$ scores tend to be highest and the $Y$ scores higher than the $Z$ scores. The $X$ scores are most variable; both their range and standard deviation are highest. The $Z$ scores exhibit the least variability. With respect to shape, the $X$ and $Y$ distributions are roughly symmetric. Note that in both instances the mean and median are nearly equal to each other and the skewness value is small relative to its standard error (see the table below). In contrast, the $Z$ distribution has a straggling right tail and is clearly skewed in that direction. A clear difference between the shapes of the $X$ and $Y$ distributions is that the former has outlying scores in both tails. Although we might expect some outliers even when scores are sampled from a normal population, 4 (20%) of 20 scores suggests that the population, though possibly symmetric, is not normally distributed. The probability plot confirms this impression. When the expected value, assuming normality, is plotted against the observed, only the $Y$ points consistently lie close to a straight line.

|               | $X$      | $Y$      | $Z$      |
|---------------|----------|----------|----------|
| N of cases    | 20       | 20       | 20       |
| Minimum       | 10.000   | 15.000   | 9.000    |
| Maximum       | 114.000  | 76.000   | 59.000   |
| Range         | 104.000  | 61.000   | 50.000   |
| Median        | 61.000   | 49.000   | 16.500   |
| Mean          | 62.700   | 49.550   | 22.700   |
| Std. Error    | 5.503    | 3.715    | 2.983    |
| Standard Dev  | 24.609   | 16.615   | 13.342   |
| Skewness(G1)  | -0.042   | -0.310   | 1.354    |
| SE Skewness   | 0.512    | 0.512    | 0.512    |
| Kurtosis(G2)  | 1.148    | -0.449   | 1.614    |
| SE Kurtosis   | 0.992    | 0.992    | 0.992    |

**2.9** Standardizing each of the three sets of scores equates their means (at 0) and standard deviations (at 1). The ranges and medians are not necessarily ordered as they were for the original three distributions. However, the shapes are the same; the skewness and kurtosis values, and their standard errors, as well as the normal probability plot are unchanged. This should serve as a reminder that standard scores are normally distributed only if the original scores are.

**2.10** The following tables summarize location and variability for the Royer 3$^{rd}$ and 4$^{th}$ grade multiplication accuracy and response time scores. We also found both box and stem-and-leaf plots to be helpful in comparing gender and grade differences.

| | Accuracy (Multacc) | | | |
|---|---|---|---|---|
| | Grade 3 | | Grade 4 | |
| | Female | Male | Female | Male |
| Mean | 74.39 | 83.18 | 91.45 | 88.32 |
| SD | 20.35 | 14.54 | 7.52 | 17.11 |
| Median | 72.22 | 84.21 | 92.86 | 93.33 |
| H-Spread | 17.68 | 18.26 | 10.05 | 10.53 |

| | Response Time (Multrt) | | | |
|---|---|---|---|---|
| | Grade 3 | | Grade 4 | |
| | Female | Male | Female | Male |
| Mean | 5.05 | 5.10 | 3.58 | 3.76 |
| SD | 2.59 | 2.51 | 1.93 | 2.69 |
| Median | 4.24 | 4.48 | 3.51 | 2.98 |
| H-Spread | 2.65 | 4.50 | 2.49 | 2.57 |

Average male accuracy scores, whether reflected in means or medians, are clearly higher in fourth grade. Standard deviations (SD) indicate that the girls' accuracy scores in the third grade are more variable but a consideration of density plots (box plots or stem-and-leaf plots) indicates that this is due to two very low outlying scores. The fact that the H-spreads are very similar confirms this impression. In the 4th grade, both distributions shift upward but the girls' average performance matches that of the boys. The girls' mean accuracy is actually somewhat higher but this is apparently due to the presence of two very low outliers in the boys data set. This is consistent with finding that the medians are quite similar and that the male SD is much larger than the female but the H-spreads are quite similar. With skewed distributions such as these, the median and H-spread, together with a knowledge of outliers, present a more useful summary of the data than do the mean and standard deviation.

In the third grade, male and female average times are quite similar. Medians are below means reflecting a skew to the right because of a few long reaction times. The H-spread, but not the SD, indicates that the middle 50% of male scores are more spread out. Both sexes respond more quickly in the fourth grade. The means indicate little difference between boys' and girls' averages but the male median is about a half second faster. The discrepancy between means and medians makes sense when we note two high outliers in the male distribution. This also contributes to the higher standard deviation. Box plots clearly show that the male times are generally faster.

**2.11** (a)

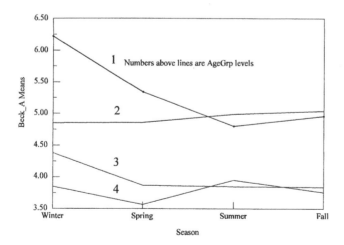

The line graph is preferable in that it more clearly reveals differences among age groups in trends over seasons.

(b) Two aspects of the graph are notable. First, the younger age groups (Agegrp = 1 and 2) have higher mean Beck anxiety scores than the older groups. Second, this is particularly pronounced in the winter season; although three of the four groups are most anxious then, this is markedly so for the youngest group.

(c) Median trends over seasons within each age group show a trend similar to that for the means, though the differences among age groups are not quite as large when the median is viewed instead of the mean.

**2.12**

(a) Both mean and median depression scores increase noticeably as Sayhlth scores increase (higher Sayhlth scores indicate poorer self-rating of health). Although we have not performed a significance test, the size of the differences and the large numbers of scores suggest that the effect will hold for other samples from the same population.

(b) Winter depression means and medians (Beck_D1 scores) are highest in categories 1 - 3. However, individuals who rate themselves in fair health (sayhlth = 4) have a somewhat higher average score in the fall season. This is particularly noticeable in the median scores.

**2.13**

(a) The $z$ score for test 1 is $z_1 = (41 - 38.6) / 4.616 = .520,$ whereas $z_2 = (51 - 46.84) / 9.496 = .438$. Performance has declined in standard deviation units.

(b) A score of 52 is the lowest integer value that transforms the test 2 score into a $z$ score exceeding .52. We arrive at this by solving $(X - 46.84)/9.496 > .52$. One point more on Test 2 would have yielded a $z$ score of .543.

(c) The almost identical values of means and medians on both tests suggests that the distributions are symmetric. This is confirmed by obtaining box plots. Finally, normal probability plots indicate that the points lie fairly close to a straight line. A few of the upper and lower points depart slightly from a straight line but this might occur by chance in any sample drawn from a normal population.

## Chapter 3

**3.1** (b) $r = .620$; (c) $\hat{Y} = 3 + 2X$; (d) $r^2 = .385$; (e) $\hat{X} = 1.577 + 0.192Y$; (f) $r^2 = .385$.

**3.2** The correlation matrix is

|   | W | X | Y |
|---|---|---|---|
| W | 1.000 | | |
| X | .374 | 1.000 | |
| Y | .927 | .482 | 1.000 |

The magnitude of a correlation is unchanged if the variables undergo linear transformations. To standardize a variable, we perform a linear transformation – we multiply it by a constant (the reciprocal of the standard deviation), then subtract a constant (the mean divided by the standard deviation)

**3.3**

(a) $r = -.487$

(b)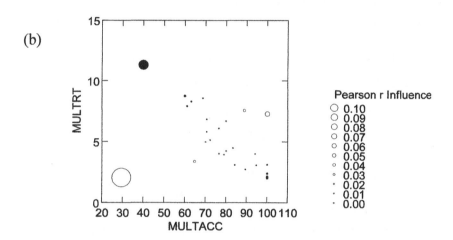

(c) If the most influential point (case 64) is removed, $r = -.743$.

**3.4**

(a) Using $r = b_1 s_X / s_Y$, we find that the correlations are .15 and .69 for situations 1 and 2, respectively – even though the slope is much higher in situation 1.

(b), (c), and (d) These transformations all leave the $z$ scores, and therefore the correlations, unchanged.

**3.5**

(a) The reasoning of the committee member is that because there is a high correlation between the pretest and posttest scores, no change in IQ has occurred. This reasoning is silly; the correlation is sensitive to the relative standing on the two tests but not the absolute scores. The correlation says nothing about the means. For example, if all the students had their scores increase by about 20 points, the correlation would be very high.

(b) Of course, the longer you live the more time you have to smoke cigarettes. If we are concerned about the influence of smoking on longevity, we should look at the rate of cigarette smoking (cigarettes/day) not the total number.

(c) The data do not allow us to make a causal statement. It could be that less able or motivated students spent less time on schoolwork and therefore had more time to watch TV. We cannot conclude that the TV watching caused the poor performance.

**3.6**

(a) Because $b_1 = rs_Y / s_X$, for men, $b_1 = (.333)\sqrt{324/100} = .599$. For women, $b_1 = (.235)\sqrt{289/25} = .799$. The slope of the regression equation for predicting income from years of experience is greater for women even though the correlation is smaller. This is because the correlation reflects not only the slope, but also the variability in $X$ and $Y$ – and the variance of $Y$ is much smaller for women than for men.

(b) To predict income from years of experience, we can use $\hat{Y} = \overline{Y} + b_1(X - \overline{X})$. For 10 years of experience, we predict for women, $\hat{Y} = 76 + .799(10 - 10) = 76.0$ and for men, $\hat{Y} = 80 + .599(10 - 15) = 77.0$. For 20 years of experience, the predictions are 84.0 for women and 83.0 for men.

**3.7** The overall correlation between height and weight is .529; however, it is only .288 for both men and women considered separately. As we discuss in more detail in Chapter 18, statistics of combined distributions may not describe any of the constituent distributions. Here, because men tend to be both taller and heavier than women, the variance of both height and weight is greater for the combined male and female distributions than for men and women considered separately:

|  | Men | Women | Overall |
|---|---|---|---|
| mean height (cm) | 176.24 | 161.43 | 169.08 |
| sd height | 6.81 | 6.73 | 10.03 |
| mean weight (kg) | 86.21 | 69.55 | 75.12 |
| sd weight | 13.81 | 16.36 | 17.24 |

**3.8** To show that $SS_{error} = \sum(Y - \hat{Y})^2 = (1 - r^2)SS_Y = SS_Y - b_1^2 SS_X$, we first note that $\hat{Y} = \overline{Y} + b_1(X - \overline{X})$ or $\hat{Y} - \overline{Y} = b_1(X - \overline{X})$. Substituting $b_1(X - \overline{X})$ for $Y - \hat{Y}$ in the expression for $SS_{error}$, we have

$$SS_{\text{error}} = \sum \left[ (Y_i - \bar{Y}) - b_1(X_i - \bar{X}) \right]^2$$

$$= \sum \left[ (Y_i - \bar{Y})^2 + b_1^2(X_i - \bar{X})^2 - 2b_1(Y_i - \bar{Y})(X_i - \bar{X}) \right]$$

Now, $\sum (Y_i - \bar{Y})^2 = SS_Y$

$$\sum b_1^2(X_i - \bar{X})^2 = b_1^2 SS_X$$

and

$$\sum \left[ -2b_1(Y_i - \bar{Y})(X_i - \bar{X}) \right] = -2b_1^2 SS_X$$

because the covariance, $s_{XY} = \dfrac{1}{N-1} \sum (X_i - \bar{X})(Y_i - \bar{Y}) = rs_X s_Y = b_1 s_X^2$,

$\sum (Y_i - \bar{Y})(X_i - \bar{X}) = (N-1)b_1 s_X^2 = b_1 SS_X$. Therefore,

$$SS_{\text{error}} = SS_Y + b_1^2 SS_X - 2b_1^2 SS_X = SS_Y - b_1^2 SS_X = SS_Y(1 - r^2)$$

because $b_1^2 SS_X = r^2 SS_Y$.

**3.9** Given that $r_{XY} = .60$, we find the following:

(a) The correlation between $Y$ and $\hat{Y}$ must also be .60 because $\hat{Y}$ is just a linear transformation of $X$.

(b) The correlation between $Y$ and $\hat{Y}$ is $corr(Y, Y - \hat{Y}) = \dfrac{1}{N-1} \sum z_Y z_{Y-\hat{Y}}$.

Note that $z_Y = \dfrac{Y - \bar{Y}}{s_Y}$ and consider the expression $z_{Y-\hat{Y}} = \dfrac{(Y - \hat{Y}) - \overline{(Y - \hat{Y})}}{s_{Y-\hat{Y}}}$.

We can see that $\overline{Y - \hat{Y}} = \dfrac{1}{N} \sum \left[ (Y - \bar{Y}) - b_1(X - \bar{X}) \right] = 0$ because deviations from the

mean sum to 0; also, $Y - \hat{Y} = (Y - \bar{Y}) - b_1(X - \bar{X}) = (Y - \bar{Y}) - (X - \bar{X})rs_Y / s_X$

and $s_{Y-\hat{Y}}^2 = \dfrac{1}{N-1} \sum (Y - \hat{Y})^2 = \dfrac{1}{N-1} SS_{\text{error}} = \dfrac{1}{N-1} SS_Y(1 - r^2) = s_Y^2(1 - r^2)$.

Therefore, $z_{Y-\hat{Y}} = \dfrac{(Y - \bar{Y}) - (X - \bar{X})rs_Y / s_X}{s_Y \sqrt{1 - r^2}}$

and

$$corr(Y, Y - \hat{Y}) =$$

$$\frac{1}{N-1} \sum z_Y z_{Y-\hat{Y}} = \frac{1}{N-1} \sum \left( \frac{Y - \bar{Y}}{s_Y} \right) \left( \frac{(Y - \bar{Y}) - (X - \bar{X})rs_Y / s_X}{s_Y \sqrt{1 - r^2}} \right)$$

$$= \frac{1}{N-1}\sum \frac{(Y-\bar{Y})^2 - (Y-\bar{Y})(X-\bar{X})rs_Y/s_X}{s_Y^2\sqrt{1-r^2}}$$

$$= \frac{s_Y^2 - r^2 s_Y^2}{s_Y^2\sqrt{1-r^2}} = \frac{1-r^2}{\sqrt{1-r^2}} = \sqrt{1-r^2}$$

Therefore, $corr(Y, Y-\hat{Y}) = \sqrt{1-r^2} = .8$

(c) $corr(\hat{Y}, Y-\hat{Y}) = \frac{1}{N-1}\sum z_{\hat{Y}} z_{Y-\hat{Y}}$

Note that $z_{\hat{Y}} = z_X$ because $\hat{Y}$ is a linear function of $X$;

From part b, $z_{Y-\hat{Y}} = \frac{(Y-\bar{Y}) - (X-\bar{X})rs_Y/s_X}{s_Y\sqrt{1-r^2}}$

so that $corr(\hat{Y}, Y-\hat{Y}) = \frac{1}{N-1}\sum\left(\frac{X-\bar{X}}{s_X}\right)\left(\frac{(Y-\bar{Y}) - (X-\bar{X})rs_Y/s_X}{s_Y\sqrt{1-r^2}}\right)$

$$= \frac{rs_X s_Y - rs_X s_Y}{s_X s_Y\sqrt{1-r^2}} = 0$$

## Chapter 4

**4.1** (a) .2; (b) $.2^3 = .008$; (c) $(.8)(.2)(.8) = .128$; (d) $(3)(.128) = .384$;
(e) $1 - p$(none correct) $= 1 - .8^3 = 1 - .512 = .488$; (f) $(3)(.2^2)(.8) = .096$;
(g) $(.8^4)(.2) = .08192$.

**4.2**
(a) (i) "yes" and "no" exhaust the possible set of responses; they are mutually exclusive because the same individual cannot have both traits; and they are not independent because the occurrence of one excludes the occurrence of the other.
(ii) "blue" and "brown" are not exhaustive (there is a third category); they are mutually exclusive; and they are not independent.
(iii) "yes" and "brown" are not exhaustive; they are not mutually exclusive (because both the trait and brown eyes can be present); they are not independent because $p$(brown|yes) $\neq p$(brown|no) (30/120 $\neq$ 110/180; alternatively, $p$(brown) $x$ $p$(yes) $\neq p$(brown *and* yes) (140/300 x 120/300 $\neq$ 30/300).
(b) (i) 70/120; (ii) 70/90; (iii) p(yes) + p(blue) - p(yes *and* blue) = 120/300 + 90/300 - 70/300 = 140/300; (iv) 70/300.
(c) 30/300 = .1; (d) .1 x .1 = .01; (e) 30/300 x 29/299.

**4.3**

(a)

| Test Results | HIV | No HIV | Total |
|---|---|---|---|
| Positive | 997 | 1,485 | 2,482 |
| Negative | 3 | 97,515 | 97,518 |
| Total | 1,000 | 99,000 | 100,000 |

(b) 997/2,482 = .402; (c) 97,515/97,518 = .99997

**4.4**

(a) p(for|male) = (70 + 10)/(80 + 40) = .67
(b) p(female *and* no opinion) = (10 + 0)/200 = .05
(c) p(no opinion|female) = (10 + 0)/(60 + 20) = .125
(d) p(no opinion|student) = (5 + 10)/(80 + 60) = .107
(e) p(male|no opinion) = 5/15 = .333

**4.5**

(a) p(reject|true) = $\alpha$ = .05
(b) p(don't reject|false) = $\beta$ = 1 - power = .2
(c) This can be solved using Bayes' theorem (see Appendix 4.3):

$$p(\text{true}|\text{nonreject}) = \frac{p(\text{true and nonreject})}{p(\text{reject})}$$

$$= \frac{p(\text{nonreject}|\text{true}) \; x \; p(\text{true})}{p(\text{nonreject}|\text{true}) \; x \; p(\text{true}) \; + \; p(\text{nonreject}|\text{false}) \; x \; p(\text{false})}$$

$$= \frac{(1-\alpha)(.3)}{(1-\alpha)(.3) + (\beta)(.7)} = .67$$

A simpler, and more transparent, approach is to assume 1,000 null hypotheses are tested. Then from the information given, we have the entries without parentheses in the following table; the remaining entries (in parentheses) are obtained by addition and subtraction:

| Hypothesis | Reject | Nonreject | Total |
|---|---|---|---|
| True | 15 | (285) | 300 |
| False | 560 | (140) | (700) |
| Total | (575) | (425) | 1,000 |

The answer is p(true| nonreject) = 285/425 = .67.

(d) From the preceding table $p$(reject) = 575/1000 = .575. The result can also be obtained as $p$(reject) = $p$(reject|true) x $p$(true) + $p$(reject|false) x $p$(false) = (.05)(.3) + (.8)(.7) = .575.

**4.6**

(a) $p$(negative|dementia) = 1 - $p$(positive|dementia) = .83.

(b) $p$(negative|no dementia) = 1- $p$(positive|no dementia) = .992.

(c)
$$p(\text{dem}|\text{pos}) = \frac{p(\text{pos}|\text{dem}) \; x \; p(\text{dem})}{p(\text{pos}|\text{dem}) \; x \; p(\text{dem}) + p(\text{pos}|\text{no dem}) \; x \; p(\text{no dem})}$$

$$= \frac{(.17)(.2)}{(.17)(.2) + (.008)(.8)} = .84$$

Alternatively, assuming 10,000 cases and constructing a table as in the answer to Exercise 4.5, $p$(dementia|positive) = 340/(340+64) = .84. Note that this is the 0 ratio provided by Bayes' theorem, with numerator and denominator multiplied by 10,000.

(d)
$$p(\text{dem}|\text{neg}) = \frac{p(\text{neg}|\text{dem}) \; x \; p(\text{dem})}{p(\text{neg}|\text{dem}) \; x \; p(\text{dem}) + p(\text{neg}|\text{no dem}) \; x \; p(\text{no dem})}$$

$$= \frac{(.83)(.2)}{(.83)(.2) + (.992)(.8)} = .17$$

Alternatively, using a table based on 10,000 cases, $p$(dem|neg) = 1,660 /(1,660 + 7,936) = .17.

**4.7**

(a) $H_0$: $p$ = .25; $H_1$: $p$ > .25; (b) $H_0$: $p$ = .20; $H_1$: $p$ > .20, although those espousing ESP often take below-chance performance as evidence for their cause as well – in which case, $H_1$: $p \neq .20$; (c) $H_0$: $p$ = .60; $H_1$: $p$ > .60.

**4.8** Let $C$ = the critical region. (a) $C > 10$; (b) $C = 5$; (c) $C < 3$; (d) $C < 5$ or $C > 16$.

**4.9**

(a) No. $p = p$(data at least as extreme as that obtained | $H_0$ true), not $p(H_0$ true | data).

(b) No, $p(H_1$ true) = .997 implies $p(H_0$ true) = .003. We can't conclude this for the same reason given in part (a).

**4.10**

(a) (i) $H_0$: $p = .25$, $H_1$: $p > .25$; (ii) $n = 5$; (iii) reject if $r$ (number matches) > 3. (b) (i) $H_0$: $p = .4$, $H_1$: $p \neq .4$; (ii) $n = 15$; (iii) reject if $r < 2$ or $r > 9$.

**4.11**

(a) Reject if $r < 7$. $p(reject) = .037 + .015 + .005 + .001 + .000 = .058$.

(b) Power = $p$(in rejection region|$\pi = .35$) = $.171 + .127 + .074 + .032 + .01 + .002 + .000 = .42$.

(c) Reject if $r > 14$ or $r < 6$.

(d) Power now equals .25.

**4.12**

(a) $H_0$: $p = .5$, $H_1$: $p > .5$

(b) Eight participants improved from the pretest to the posttest. For $n = 10$, $p = .5$, and $\alpha = .06$, the rejection region is $r \geq 8$; $p(r \geq 8) = .055$. Therefore, the null hypothesis is rejected.

(c) The reasoning is that if the observed probability of a success (improved score) does not differ significantly from .75, the training method is a success. However, failing to obtain a significant result is not the same as accepting $H_0$. A failure to achieve a significant result may reflect too little power, perhaps due to too small an $n$.

**4.13**

(a) $H_0$: $p = .5$, $H_1$: $p \neq .5$; $p$ is the probability that the imagery is better than the rote procedure. Let $n = 12$ and $\alpha = .05$. Then the decision rule is: reject if $r > 9$ or $r < 3$ where $r$ is the number of participants who performed better using the imagery method. The null hypothesis cannot be rejected.

(b) Power = $p(r < 3$ or $> 9 | \pi = .9) = p(r < 3$ or $> 9 | \pi = .1) = .282 + .377 + .230 + 0 = .89$.

**4.14** (a) In answering Exercise 4.10, part (a), we found the rejection region to be $r > 3$. For $n = 5$, power = $p$(reject|$\pi = .5$) = $.156 + .031 = .19$. Clearly, the $n$ was too small to provide a good test of the null hypothesis.

(b) By "true probability" we mean that $\pi$ is the proportion of the population sampled that would exhibit empathy in this study. Alternatively, if the study were replicated many times, the average proportion of empathic participants should approach the "true probability."

**4.15**

(a) $E(Y) = \sum_{y} y \cdot p(y) = (.25)(20) = 5;$

$$var(Y) = E[y - E(y)]^2$$
$$= E(y^2) - [E(y)]^2$$
$$= (.25)(120) - 5^2 = 5$$

(b) The completed sampling distribution is

| $\overline{Y}$ = | 2 | 3 | 4 | 5 | 6 | 7 | 8 |
|---|---|---|---|---|---|---|---|
| $p(\overline{Y})$ = | .0625 | .125 | .1875 | .250 | .1875 | .125 | .0625 |

The $p(\overline{Y})$ are obtained by enumerating the different ways a value of $\overline{Y}$ can occur and then multiplying $.25^2$ by the number of ways. For example,

$p(\overline{Y}) = 5 = p(Y_1 = 2$ and $Y_2 = 8) + p(Y_1 = 8$ and $Y_2 = 2) + p(Y_1 = 4$ and $Y_2 = 6) + p(Y_1 = 6$ and $Y_2 = 4) = (4)(.25^2) = .25;$ the subscripts refer to the first and second scores drawn from the sample.

(c) $E(\overline{Y}) = (.0625)(1) + ... + (.0625)(8) = 5 = E(Y)$

$var(\overline{Y}) = [(.0625)(1^2) + ... + (.0625)(8^2)] - 5^2 = 2.5 = var(Y) / 2;$ in general, the variance of the sample mean equals the sample variance divided by the sample size.

**4.16**

(a) $E(Y) = \sum_{y} y \cdot p(y) = (.2)(30) = 6;$

$$var(Y) = E[y - E(y)]^2$$
$$= E(y^2) - [E(y)]^2$$
$$= (.2)(220) - 6^2 = 8$$

(b) The completed sampling distribution is

| $\overline{Y}$ = | 2 | 3 | 4 | 5 | 6 | 7 | 8 | 9 | 10 |
|---|---|---|---|---|---|---|---|---|---|
| $p(\overline{Y})$ = | .04 | .08 | .12 | .16 | .20 | .16 | .12 | .08 | .04 |

For example, $\overline{Y} = 4$ can be obtained by sampling either <2,6>, <6,2>, or <4,4>. Each of these three samples occurs with probability $(1/5)^2 = .04$.

(c) $E(\overline{Y}) = (.04)(2) + ... + (.04)(10) = 6 = E(Y)$;

$var(\overline{Y}) = [(.04)(2^2) + ... + (.04)(10^2)] - 6^2 = 4 = var(Y)/2$

## 4.17

(a) Because the school has several thousand students, removing one score of 150 should have little effect on the mean of the remaining population. For example, if there are 1,000 students, the mean of the remaining 999 students is 99.95. Therefore, the best estimate of the mean of the other four students is still 100.

(b) Given a score of 150 for the first student, and an estimate of 100 for the mean of the remaining four students, our best estimate of the sample mean is 550/5 = 110.

(c) Only the answer to (b) is changed. As sample size increases, the estimate of the sample mean more closely approximates the population mean. In this case, the estimate is [150 + (9)(100)]/10 = 105.

## 4.18

(a)

| $\overline{Y}$ = | 3 | 4 | 5 | 6 | 7 | 8 | 9 |
|---|---|---|---|---|---|---|---|
| $p(\overline{Y})$ = | .1 | .1 | .2 | .2 | .2 | .1 | .1 |

For example, $\overline{Y}$ = 5 occurs with samples <2,8>, <8,2>, <4,6>, and <6,4>. Each of these samples has $p = (1/5) \times (1/4) = .05$. Therefore, $p(\overline{Y}=5)=4 \times .05=.2$

(b) $E(\overline{Y})= E(Y)=6$; $var(\overline{Y}) = \sum p(\overline{Y}) \cdot (\overline{Y}-6)^2 = 3$. Alternatively, the variance of the sampling distribution can be calculated as

$$var(\overline{Y}) = \sum p(\overline{Y}) \cdot (\overline{Y})^2 - [E(\overline{Y})]^2 = 3.$$ Note that the variance of the mean is not

$\sigma^2/n$ when sampling is without replacement. In the present example, the mean cannot take the values 2 and 10 and density is more concentrated in the values near the mean than when sampling was with replacement; this reduces the variance relative to the result when sampling was with replacement (Exercise 4.16).

## Chapter 5

**5.1** The answers to this exercise and several others involve calculating $z$ scores and then using Appendix Table C.2. In some cases, it may be useful to draw a normal curve, shading the area asked for.

(a) (i) $z = (130 - 100)/15 = 2$; $p(z > 2) = .023$. (ii) $z_{upper} = (145 - 100)/15 = 3$; $z_{lower} = (85 - 100)/15 = -1$; $p = (.5 - .001) + (.5 - .159) = .840$; (iii) $z = (70 - 100)/15 = -2$. Therefore, $p = .5 + (.5 - .023) = .977$; (iv) $z_{70} = -2$,

$z_{80}$ = -1.33. $p$ = .092 - .023 = .069.

(b) $z$ = -1.28 and $z$ = 1.28 contain the middle .80 of the area under the normal curve. Therefore, $1.28 = (Y_{upper} - 100)/15$. Solving algebraically, $Y_{upper}$ = 119.2. Similarly, $Y_{lower}$ = 80.8.

(c) $z_{.75}$ = .675 = $(Y_{.75} - 100)/15$. Solving, $Y_{.75}$ = 110.125.

(d) $z$ = 1.00; therefore, $p$ = .159.

(e) The appropriate denominator of the $z$ is the standard error of the mean, Therefore, $z = (115 - 100)/(15/\sqrt{10}) = 3.16$. The area above this value is essentially zero.

### 5.2

(a) $z$ = (170 - 200)/60 = -.5; $p$ = .5 + (.5 -.309) = .691.

(b) $z = (170 - 200)/(60/\sqrt{9}) = -1.5$; $p$ = .5 + (.5 -.067) = .933. Note the use of the standard error rather than the standard deviation.

(c) Using $M$ for men and $W$ for women, we have $E(D) = E(W - M) = E(W) - E(M) = 30$. $\sigma_D = \sqrt{(\sigma_W^2 + \sigma_M^2)} = 78.1025$.

(d) We want $p(W > M) = p(W - M > 0)$. Then $z_0 = (0 - 30)/78.103 = -.384$ and $p(z > -.384) = .650$.

### 5.3

(a) (i) $z$ = -.25; therefore, $p$ = .401. (ii) $z$ = 1.5; therefore, $p$ = .067. (iii) $z_{lower} = -.75$; the area above this is .773. $z_{upper} = .5$; the area above this is .309. The area between the two $z$ scores is $p$ = .773 - .309 = .464.

(b) To find $p(X > \mu_Y)$, calculate $z = (\mu_Y - \mu_X)/\sigma_X$ = (20 - 30)/20 = -.5. The proportion of the area in the $X$ distribution above 20 ($\mu_Y$) is $p$ = .69.

(c) We require the population mean and variance of $W$. $\mu_W = \mu_X + \mu_Y = 50$. $\sigma_W = \sqrt{\sigma_X^2 + \sigma_Y^2} = = 25.61$. Because $X$ and $Y$ are normally distributed, $W$ is. Therefore, to get $p(W > 35)$, calculate $z$ = (35-50)/25.61 = -.59. Then, $p$ = .72.

(d) This requires finding the values of $X$ and $Y$. From the information given, $z_X$ = 1.035 = $(X - 30)/20$. Solving, $X$ = 50.70. Similarly, $z_Y$ = -.52 = $(Y - 20)/16$ and $Y$ = 11.68. Therefore, $W$ = 62.28, $z_w$ = (62.28 - 50)/25.61 = .48, which exceeds approximately .68 of the $W$ population.

### 5.4

(a) (i) $p(Y < .6) = .6$; (ii) $p = .6^2$; (iii) $p = .6^{20}$.

(b) The sample will be symmetric and approximately normal. The mean will be the same as the mean of the original population, .5, and the standard deviation will be $\sigma / \sqrt{20} = .065$.

(c) $z = (.6 - .5)/.065 = 1.54$. The area below this $z$ score is $1 - .06 = .94$.

(d) The Central Limit theorem states that as $N$ increases, the sampling distribution of the mean approaches normality, justifying our approach to part (c). This approach is not correct for part (a), (iii) because there the distribution in question was the population distribution, which was not normal.

## 5.5

(a) $H_0$: $\pi = .4$; $H_1$: $\pi < .4$.

(b) $\sigma_p = \sqrt{(.4)(.6) / 48} = .0707$. Then $z = [(12/48) - .4]/.0707 = -2.12$. The decision rule is: reject $H_0$ if $z < -1.645$. Therefore, reject the null hypothesis. We conclude that the new drug has reduced the probability of reoccurrence of symptoms.

(c) (i) We assumed that the sampling distribution of $p$ was normal. (ii) Because $p$ is an average of 48 1's (failures) and 0's (successes), the central limit theorem provides the rationale for using the normal distribution. (iii) Because the distribution of $p$ approaches normality as $N$ increases, the normal approximation will be less good when $N = 10$ than when $N = 48$.

## 5.6

(a) $H_0$: $\mu = 52.8$; $H_1$: $\mu > 52.8$. Reject if $z > 1.645$.

(b) $\sigma_{\bar{X}} = 10.5/\sqrt{50} = 1.485$. Therefore, $z = (56 - 52.8)/1.485 = 2.15$. Reject $H_0$ and conclude that the population mean of authoritarianism scores has increased.

(c) The mean of 57 is 2.83 $\sigma_{\bar{X}}$ units above 52.8. Power is the area to the right of 1.645 under the distribution with mean at 2.83. Therefore, calculate $z = (1.645 - 2.83)/1.485 = -1.18$; power = the area above this $z$ score, or $1 - .119 = .88$.

(d) $CI = 56 \pm (1.96)(1.485) = 53.09, 58.91$.

(e) Although the one interval calculated either contains $\mu$ or doesn't, we have .95 confidence in the following sense: If we were to draw many samples of size $n$ from a normally distributed population with the assumed variance, .95 of the intervals calculated would contain the true mean.

## 5.7

(a) (i) $E(X) = .2$; $var(X) = \pi(1 - \pi) = .16$. (ii) $var(\bar{X}) = .16/3 = .053$.

(b)

| $Y$ | $p(Y)$ | $\overline{X}$ | $S_X^2$ | $S_{\overline{X}}^2$ | $s_{\overline{X}}^2$ |
|---|---|---|---|---|---|
| 0 | $.8^3 = .512$ | 0 | 0 | 0 | 0 |
| 1 | $(3)(.8^2)(.2) = .384$ | 1/3 | 2/9 | 2/27 | 1/9 |
| 2 | $(3)(.8)(.2^2) = .096$ | 2/3 | 2/9 | 2/27 | 1/9 |
| 3 | $.2^3 = .008$ | 1 | 0 | 0 | 0 |

When $\overline{X} = 0$ or 1, the sample variance must be 0 because all scores must be the same. To calculate $S_{\overline{X}}^2$ when $Y = 1$, note that the three scores must be 0, 0, and 1. Then $S_X^2 = [(1 - 1/3)^2 + (2)(0 - 1/3)^2]/3 = 2/9$ and $s_X^2 = (3/2)(2/9) = 1/3$; therefore, $S_{\overline{X}}^2 = 2/27$ and $s_{\overline{X}}^2 = 1/9$. Similarly, when $Y = 2$, $S_{\overline{X}}^2 = 2/27$ and $s_{\overline{X}}^2 = 1/9$.

(c) (i) $E(Y) = (.384)(1) + (.096)(2) + (.008)(3) = .6$; (ii) $E(\overline{X}) = .2$; (iii) $E(S_{\overline{X}}^2) = 0 + (.384 + .096)(2/27) + 0 = .036$; (iv) $E(s_{\overline{X}}^2) = 0 + (.384 + .096)(1/9) + 0 = .053$.

(d) $E(Y) = N \times E(X)$ and $E(\overline{X}) = E(X)$.

(e) $E(S_{\overline{X}}^2) = [(N - 1)/N] \times var(\overline{X})$; $E(s_{\overline{X}}^2) = var(\overline{X})$.

**5.8**

(a) $E(UM) = (1/2)E(\overline{Y}_1 + \overline{Y}_2) = (1/2)(2 \cdot \mu) = \mu$; UM is an unbiased estimator.

(b) $E(WM) = E\left[\left(\dfrac{n_1}{n_1 + n_2}\right)\overline{Y}_1 + \left(\dfrac{n_2}{n_1 + n_2}\right)\overline{Y}_2\right] = \left(\dfrac{n_1 + n_2}{n_1 + n_2}\right)\mu = \mu$

WM is also an unbiased estimator.

(c) $var(w_1\overline{Y}_1 + w_2\overline{Y}_2) = w_1^2(\sigma^2/n_1 + \sigma^2/n_2)$ where $w_1$ and $w_2$ are weights. For UM, both weights $= 1/2$. Substituting the values given, $var(UM) = (1/2)^2(4/20 + 4/80) = .0625$.

(d) $var(WM) = (1/5)^2(4/20) + (4/5)^2(4/80) = .04$.

(e) Although both UM and WM are unbiased estimators of $\mu$, WM has the smaller sampling variance (squared standard error) and is therefore the better estimator because its values are more likely to be close to the parameter being estimated.

(f) Of the two means, $var(\bar{Y}_2)$ is based on the larger sample and will have the smaller standard error. Its variance is $4/80 = .05$. WM is a more efficient estimator.

**5.9**

(a) $E(T-C) = \mu_T - \mu_C = .5\sigma$ ; $var(T-C) = \sigma_T^2 + \sigma_C^2 = 2\sigma^2$.

(b) $p(T > C) = p(T - C > 0)$. Therefore, get

$z = [0 - E(T - C)\sqrt{var(T - C)} = (0 - .5\sigma)/\sigma\sqrt{2} = -.35; p\,(z > -.35) = .64$.

(c) (i) $z = \dfrac{0 - .5\sigma}{(\sigma/3)\sqrt{2}} = -1.06 ; p = .86$. (ii) $z = \dfrac{0 - .2\sigma}{(\sigma/3)\sqrt{2}} = -.42 ; p = .66$.

This change in $p(T > C)$ with changing effect size corresponds to changes in power of a significance test of the difference. Similarly, if $N$ were increased, $p$ would increase because the denominator of the $z$ (the standard error) would decrease. The power of a significance test would also increase.

**5.10**

(a) Let $D$ = democrat, $L$ = liberal, and $C$ = conservative. Then from the developments in Chapter 4,

$$p(D) = p(D \text{ and } L) + p(D \text{ and } C)$$
$$= p(D|L) \cdot p(L) + p(D|C) \cdot p(C)$$
$$= (.9)(1/2) + (.3)(1/2) = .6$$

(b) $var(p_D) = (.6)(.4)/50 = .0048$

(c) (i) $var(p_{D|L}) = (.9)(.1)/25 = .0036$; (ii) $var(p_{D|C}) = (.3)(.7)/25 = .0084$;

(iii)
$$var(p_D | \text{stratification}) = var[(1/2)(p_{D|L} + p_{D|C})]$$
$$= (1/2)^2(\sigma_{D|L}^2 + \sigma_{D|C}^2)$$
$$= (.25)(.0036 + .0084) = .0030.$$

(iv) The estimate of $p_D$ based on stratification is more efficient; that is, there is less sampling variance.

**5.11**

(a) Let $\bar{D} = \bar{Y}_1 - \bar{Y}_2$. Then $s_{\bar{D}} = \sqrt{(1/n)(s_1^2 + s_2^2 - 2rs_1s_2)} = 1.469$.

(b) $CI = \bar{D} \pm s_{\bar{D}} z_{.025}$

$\qquad = 5.241 \pm (1.469)(1.96) = 2.36, \ 8.12$

(c) $t = \bar{D}/s_{\bar{D}} = 5.241/1.469 = 3.57$. Total cholesterol levels are significantly higher in the winter than in the spring season. Note that this result is consistent with the fact that 0 was not contained within the .95 *CI* calculated in part (b). The higher TC values in winter may reflect less exercise or diets that are more restricted to high cholesterol items (e.g., fewer vegetables in winter).

(d) SPSS yields confidence bounds of 2.347 and 8.135, and a value of *t* of 3.569, all values that are close to those obtained using the *z* test.

### 5.12

(a) The equation for the interval width is

$$[\bar{Y} + (20/\sqrt{n})(1.96)] - [\bar{Y} - (20/\sqrt{n})(1.96)] = 4 .$$

Solving, $n = [(2)(20)(1.96)/4]^2 = 384$ subjects.

(b) The distance between the null and alternative distributions is $\delta/\sigma_D$ where $\delta$ is $\mu_1$ - $\mu_2$, and $\sigma_D$ is the standard deviation of the population of difference scores. We want to know the power to reject the null hypothesis when $\delta/\sigma_D =$ .24, $N = 100$, and $\alpha = .05$. Assuming the specific alternative, the sampling distribution of the mean difference has its mean at $\delta/(\sigma/\sqrt{n}) = (\delta\sqrt{n})/\sigma = 2.4$. The critical value under the null hypothesis is 1.645. Power is the area to the right of the critical value under the alternative hypothesis. Therefore, we want the probability of exceeding 1.645 - 2.4 = -.755 under the normal curve. This is approximately .78.

### 5.13

(a) The standard error of the sampling distribution of the difference between the means is $s_{\bar{D}} = \sqrt{(41.898)^2/112 + (33.896)^2/98} = 5.234$ . Therefore, the *CI* = (228.931 - 215.455) $\pm$ (5.234)(1.96) = 3.217, 23.735.

(b) $s_{\bar{D}} = \sqrt{(2)(40^2)/100} = 5.657$. The mean of the alternative distribution is at $z = 15/5.657 = 2.65$. The critical *z* value, 1.645, is -1.005 with respect to this mean. Therefore, power is p($z > -1.005$) = .84.

### 5.14

(a) $H_0$: $\mu = 200$; $H_1$: $\mu > 200$.
(b) $H_A$: $\mu = 206$.
(c) The *z* corresponding to 206 is $z = 6/(30/\sqrt{n}) = \sqrt{n}/5$. In order to have .8 power, this *z* must be .84 above the critical *z* of 1.645. Therefore, $z = \sqrt{n}/5 = 1.645 + .84$; $n = [(5)(2.485)]^2 = 154$.

### 5.15

(a) Consider only the left two columns of statistics ("Original Statistics") in the following table. With respect to location, the medians and means for women are higher than those for men. Depression scores are also more variable for women. Skewness and kurtosis measures are high for both sexes. We graphed histograms, box plots, normal probability curves, and stem-and-leaf plots. It is clear that the data are markedly skewed to the right in both groups with many outliers (roughly, a little more than 7 %).

| | Original Statistics | | Statistics - No Outliers | |
|---|---|---|---|---|
| | Men | Women | Men | Female |
| # of cases | 156 | 164 | 144 | 152 |
| Minimum | 0.002 | 0.000 | 0.002 | 0.000 |
| Maximum | 22.987 | 28.989 | 13.250 | 16.659 |
| Median | 3.107 | 4.375 | 2.938 | 4.157 |
| Mean | 4.868 | 6.402 | 3.834 | 5.087 |
| Standard Dev | 4.725 | 6.215 | 3.108 | 4.071 |
| Skewness(G1) | 1.605 | 1.711 | 1.005 | 1.082 |
| SE Skewness | 0.194 | 0.190 | 0.202 | 0.197 |
| Kurtosis(G2) | 2.389 | 2.812 | 0.406 | 0.521 |
| SE Kurtosis | 0.386 | 0.377 | 0.401 | 0.391 |

(b) $CI = (6.402 - 4.868) \pm 1.96\sqrt{\dfrac{4.725^2}{156} + \dfrac{6.215^2}{164}} = .333, \ 2.735$

The interval does not include zero so the null hypothesis of no difference can be rejected. Although the populations of scores are not normal, the sampling distributions of the means are likely to be, given the size of the samples, and therefore the significance test is likely to be valid.

(c) From our stem-and-leaf plots, we find that male scores greater than 13.250 and female scores greater than 16.659 are outliers.

(d) The "statistics - no outliers" column contains the results obtained when outliers, as defined in part (c), are deleted from the data set. There is still considerable skew although the skew (G1) statistics indicate that there is less than previously. Some scores that previously were not outliers are so now but there are fewer outliers than previously. Also, the mean and median are closer together in both groups than in the original full data set. The removal of outliers greatly reduces variability as indicated by the smaller variances. Consequently the new CI bounds, .430 and 2.076, are closer than those obtained in part (b). We have a more precise estimate of the population difference in means, and a more powerful test of the null hypothesis.

## Chapter 6

### 6.1

(a) The standard error of the difference is $s_{\overline{D}} = s / \sqrt{n} = 1.555$ and $\overline{D} = 4.50$. With 11 *df*, the .05 (two-tailed) critical value of *t* is 2.2016. Therefore the CI bounds are $= 4.5 \pm (2.201)(1.555) = 1.077, 7.923$.

(b) $H_0$: $\mu_D = 0$; $H_1$: $\mu_D \neq 0$; reject if $|t| > 2.201$. $t = \overline{D} / s_{\overline{D}} = 4.5/1.555$

$= 2.895$. Reject $H_0$.

(c) The $.95\ CI$ is $4.5 \pm (2.074)(5.332) = -6.656, 15.584$ and $t = 4.5/5.332 = .844$, clearly less than the critical value on 22 $df$ of 2.074. The independent-groups analysis is considerably less efficient as evidenced by the fact that the confidence interval is more than three times wider than in the repeated-measures design. The independent-groups design has twice as many degrees of freedom, but this is more than compensated for by the much smaller standard error for the repeated-measures data. That standard error is smaller because we constructed the data with a very high correlation (.93) between the *Easy* and *Hard* scores. Because the variance of the difference is the sum of the variances minus the covariance (which is a function of the correlation; see Appendix 5.1), the result is a small standard error. Although the correlation in these data is higher than would be found in most studies, the repeated-measures design will usually be more efficient. However, one potential problem is that exposure to one condition may affect performance under the other condition, and sometimes it is impossible to test subjects in two conditions (e.g., two types of surgery).

### 6.2

(a) $\overline{Y}_1 = 47.889$, $s = 14.987$, and $SEM = 4.997$. The critical $t$ value is 1.860. Therefore, $CI = 47.889 \pm (1.86)(4.997) = 38.60, 57.18$.

(b) Because the upper limit of the $.90\ CI$ is below the null hypothesis value of 60, it follows that the null hypothesis can be rejected at the .05 level in favor of the alternative that the mean motor skill score of the protein-deficient population below 60. A $t$ test yields $t = (47.889 - 60)/4.997 = -2.424$, $p = .021$.

(c) As opposed to the situation with the normal distribution, we really have to use successive approximation because as we change the value of $n$, the critical value of $t$ changes as well. Start by solving $(2)(1.86)(14.987)/\sqrt{n} = 12$. This yields $n \approx 22$. If $n = 22$, $df = 21$, so we redo the calculation, substituting $t_{crit,.05,21} = 1.721$ for 1.86, yields $n = 19$. Redoing a third time with $df = 18$ still yields $n = 19$.

(d) $\overline{Y}_2 = 50.889$, and $s = 13.365$, $SEM = 4.455$. $t_{.05} = 1.86$. $CI = 50.889 \pm (1.86)(4.455) = 42.61, 59.17$. The mean is again significantly below 60; $t = -2.405$, $p = .038$.

(e) $\overline{D} = \overline{Y}_2 - \overline{Y}_1 = 3$, $s = 3.742$, $s_{\overline{D}} = 1.247$. $CI = .68, 5.32$. $t = 2.405$, $p = .022$.

(f) $E_S = 3/3.742 = .80$. Although performance is still below the norm [part (d)], the improvement is significant and large, according to Cohen's guidelines.

### 6.3

(a) $t = 2/(5.6/4) = 1.43$. The null hypothesis cannot be rejected.

(b) $E_S = 2/5.6 = .36$.

(c) Power, based on the noncentral $t$ distribution (using either a statistical software package or the free software cited in the chapter), $= .39$ when $n = 16$.

(d) When $n = 36$, power $= .68$.

(f) The power values for various $n$s and the two distributions are:

| | Distributions | |
|---|---|---|
| $n$ | $t$ | $z$ |
| 16 | .39 | .41 |
| 36 | .68 | .69 |
| 49 | .80 | .81 |

Two points should be evident: First, the $z$ provides a reasonable approximation to the power of the $t$ test, even when $n$ is relatively small. Second, because the $t$ distribution approaches the normal with increasing degrees of freedom, the approximation improves as $n$ increases.

### 6.4

(a) The pooled variance is $s_{pool}^2 = [(17)(16) + (13)(20)]/30 = 17.733$.

The standard error of the difference of the means is $s_{\bar{D}} = \sqrt{s_{pool}^2 \left( \frac{1}{18} + \frac{1}{14} \right)} = 1.501$.

The $df = 30$; $t_{30,.025}$ 2.042 and $CI = 2.4 \pm (2.042)(1.501) = -.664, 5.464$. The null hypothesis cannot be rejected. Nor can any other hypothesized difference within this interval.

(b) $E_S = (\bar{Y}_1 - \bar{Y}_2)/s_{pool} = 2.4/4.211 = .57$. Power $= .34$ for a two-tailed test.

(c) GPOWER's *a priori* option gives $N = 100$ to get power $= .8$; therefore, $n = 50$.

(d) With equal $n$s, $s_{\bar{D}} = \sqrt{(s_1^2 + s_2^2)/n} = \sqrt{36/50} = .849$. Also, $t_{98,.025} = 1.99$. Therefore, the interval width is $(2)(1.99)(.849) = 3.379$. We have increased the precision of our estimate, reducing the interval width by almost one-half.

### 6.5

(a) $s_{pool}^2 = [(20)(8) + (10)(30)]/30 = 15.333$. The standard error of the

difference of the means is $s_{\bar{D}} = \sqrt{s_{pool}^2 \left( \frac{1}{21} + \frac{1}{11} \right)} = 1.457$. The $df = 30$; $t_{30,.025} = 2.042$. $t = 3.2/1.457 = 2.20$. Therefore, reject $H_0$.

(b) Applying Equation 6.19, we have $t' = 3.2/1.763 = 1.82$. From Equation 6.20, $df' = 13$. The null hypothesis cannot be rejected against a two-tailed alternative.

(c) The pooled-variance test gives heavy weight to the smaller variance, producing a positive bias (i.e., too many Type 1 errors) in the $t$ test. The separate-variance test corrects this bias.

**6.6**

(a) (i) $CI = (\overline{Y}_{10,G} - \overline{Y}_{8,G}) \pm (t_{18, .10})(s_{\overline{D}}) = 7 \pm 1.734\sqrt{(2.9^2 + 2.2^2)/10} =$ 5.004, 8.996. (ii) The lower limit is well above zero indicating that a one-tailed $t$ test would reject $H_0$ at the .05 level.

(b) (i) $H_0:(\mu_{10,B} - \mu_{10,G}) - (\mu_{8,B} - \mu_{8,G}) = 0$;

$H_1:(\mu_{10,B} - \mu_{10,G}) - (\mu_{8,B} - \mu_{8,G}) > 0$

The standard error of the linear combination of interest is $\sqrt{(2.7^2 + 2.1^2 + 2.9^2 + 2.2^2)/10.} = 1.580$. Therefore, $t = [(72 - 60) - (58 - 53)]/1.580 = 4.430$. The gap between boys' and girls' scores has significantly increased with age.

**6.7**

(a) $H_0: \mu_H - \mu_L = 0$; $H_1: \mu_H - \mu_L \neq 0$; for $\alpha = .05$ and $df = 34$, reject if $|t| > 2.034$. To calculate the standard error of the difference of the means ($s_{\overline{D}}$), we first obtain $s_{pool}^2 = [(14)(6.102^2) + (20)(6.128^2)]/34 = 37.421$. Multiplying by $(1/15) + (1/21)$, and taking the square root, $s_{\overline{D}} = 2.068$, and $t = (67.333 - 66.048)/2.068 = .621$ which is not significant.

(b) $H_0: \mu_M - (.5)(\mu_H + \mu_L) = 0$; $H_1: \mu_M - (.5)(\mu_H + \mu_L) > 0$; $df = N - 3 = 51$ and for $\alpha = .05$, reject $H_0$ if $t > 1.676$. $\hat{\psi} = 70.611 - (.5)(67.333 + 66.048) = 3.921$. The variance of this linear combination is $s_{\hat{\psi}}^2 = s_M^2/n_M + (.5^2)(s_L^2/n_L + s_H^2/n_H)$. However, because the group variances are quite similar, we find $s_{pool}^2 = [(n_L - 1)s_L^2 + (n_M - 1)s_M^2 + (n_H - 1)s_H^2]/(N - 3) = 37.502$. Then $s_{\hat{\psi}}^2 = s_{pool}^2[1/n_M + (.5^2)(1/n_L + 1/n_H)] = 3.155$ and, taking the square root, $s_{\hat{\psi}} = 1.776$. Finally, $t = \hat{\psi}/s_{\hat{\psi}} = 3.931/1.776 = 2.21$. The null hypothesis can be rejected.

(c) The standardized contrast is $\hat{\psi}/s_{pool}$. For the contrast in part (a), $\hat{\psi}_S = 1.285/\sqrt{37.421} = .21$. If there is an effect, it is small. For the contrast in part (b), $\hat{\psi}_S = 3.921 / \sqrt{37.502} = .64$, a considerably larger effect.

**6.8**

(a) With 7 *df*, $t_{7,.05} = 2.365$. For the *Lab* group, the standard error of the difference is $s_{\overline{D}} = s / \sqrt{n} = 6.116 / 2.828 = 2.163$; similarly, for the *Natural* group, the *SE* is 5.165. Therefore, the two confidence intervals are

*Lab*: $CI = 15.625 \pm (2.365)(2.162) = 10.51, 20.74$;
*Natural*: $CI = 21.5 + (2.365)(5.165) = 9.29, 33.71$

It is clear that in both groups performance deteriorated significantly from the first to the second test on the materials. However, the *Natural* group forgot more from the first to the second test; we will see in part (c) whether this difference between the groups is significant. Another difference between the groups is that the interval width is wider in the *Natural* group indicating somewhat less precision in estimating the mean population change score.

(b) $s_{\overline{D}} = \sqrt{(111.979 + 59.352)/8} = 4.628$; therefore,
$t = (33.375 - 25.250)/4.628 = 1.76; p = .10$. The difference is not significant.

(c) $H_0: \mu_{D\,Lab} = \mu_{D\,Natural}$ ; $H_1: \mu_{D\,Lab} \neq \mu_{D\,Natural}$ ; $s_{\overline{D}} = 5.599$; $t = 1.049$; $p = .31$. Again, there is no evidence to suggest that the groups differ. However, this may reflect a lack of power due to the small *n*.

**6.9** A table of statistics for Sayhlth 2 and 4 should aid calculations and discussion:

|  | Sayhlth = 2 | Sayhlth = 4 |
| --- | --- | --- |
| N of cases | 182 | 23 |
| Mean | 4.560 | 11.087 |
| Std. Error | 1.506 | 4.040 |
| Standard Dev | 20.318 | 19.375 |

(a) Sayhlth = 2: $E_S = 4.560/20.318 = .22$, a small standardized effect; Sayhlth = 4: $E_S = 11.087/19.375 = .57$, a medium standardized effect.

(b) Sayhlth = 2: $t_{181,.05} = 1.973$; $CI = 4.560 \pm (1.973)(1.506) = 1.59, 7.53$; $t = 4.56/1.506 = 3.030$; $p = .003$.

Sayhlth = 4: $t_{22,.05} = 2.074$; $CI = 11.087 \pm (2.074)(4.040) = 2.71, 19.47$; $t = 11.087/4.04 = 2.744$; $p = .012$.

(c) The statistics in part (b) might suggest that the increase in TC scores in the winter relative to the spring is more pronounced in the Sayhlth = 2 group. However, this probably reflects the fact that there are many more participants in the study who rated themselves in very good health (Sayhlth = 2) than who rated themselves in fair health (Sayhlth = 4). The comparison of effect sizes serves to remind us of this because the standardized effect size is considerably larger in the Sayhlth = 4 group. Because of the small $n$ in that group, we can reach no firm conclusion about the relative effects in the two groups. We need a considerably larger sample of people who rate themselves in only fair health.

**6.10** Contrary to reports of male superiority in math, there is no clear evidence of differences in the two populations. For addition, subtraction, multiplication, and an average accuracy measure, all $t$ statistics were less than 1. Confidence intervals for each of the four measures had approximately the same bounds (within 1 or 2 points for both boys and girls. Girls actually scored higher in multiplication by 1.2 points (mean) and 3.75 points (median). However, the gender effect size for the MULTACC measure was only .008; although the largest of the four effect sizes, it was very small by any standard. With respect to shape, all distributions were skewed with low outliers, and the skew values were similar across genders and the four arithmetic measures. There is some variation in kurtosis because the shapes of histograms don't match perfectly. For example, on the ADDACC, there were about 10% more boys' than girls' scores in the 95-100 interval of our histogram. Nevertheless, the similarities in the histograms and box plots were more notable than any differences. Variances were almost identical for the average accuracy score (ACC), and although ADDACC scores were less variable for girls than boys and their multiplication scores were more variable, no ratio of variances reached even 2:1. Whether these variance differences are significant, is a question we postpone to the next chapter.

**6.11** Response time data are – like the accuracy data – quite similar for boys and girls. Again, effect sizes are very small, all less than .01, and all $t$ statistics are less than 1. All distributions are skewed to the right, with the bulk of scores falling between 1 - 3 seconds. Perhaps the one notable difference between boys and girls is that the variances of the male scores are higher for all four measures. However, the ratios are less than 2 and tests of significance will have to await developments in Chapter 7.

### Chapter 7

**7.1**

(a) (i) $df = 5$. From Appendix Table C.4, $p(\chi_5^2 < 9.236) = 1 - .10 = .90$;

(ii) $p(\chi_5^2 > 1.145) = .95$ and $p(\chi_5^2 > 6.626) = .25$. Therefore,

$p(1.145 < \chi_5^2 < 6.626) = .95 - .25 = .70$.

(b) $\chi^2 = (n-1) s^2/\sigma^2 = 5s^2/10 = s^2/2$. If $s^2 < 8.703$, then $\chi^2 < 8.703/2$, or 4.351. Therefore, $p = .50$.

**7.2**

(a) $H_0$: $\mu = 64$; $H_1$: $\mu > 64$.  $t_{30} = (66 - 64)/\sqrt{14/31} = 2.98$. There is a significant improvement.

(b) $H_0$: $\sigma^2 = 10$; $H_1$: $\sigma^2 > 10$.  $\chi^2 = (30)(14)/10 = 42$. On 30 $df$, the value of $\chi^2$ exceeded with $p = .05$ is 43.773, so we cannot conclude that the variance is greater than under the old method of teaching arithmetic.

(c) We need to apply Equation 7.5. This, in turn, requires the .05 and .95 values of $\chi^2$ on 30 $df$. These values are 43.773 and 18.493. Then the .90 limits are (30)(14)/43.773 and (30)(14)/18.493, or 9.59 and 22.71.

**7.3**

(a) $H_0$: $\sigma^2 = 12.64$; $H_1$: $\sigma^2 < 12.64$.  $\chi^2 = (9)(3.51)/12.64 = 2.499$. $p(\chi_9^2 < 2.499) = .019$. Therefore, reject $H_0$; the variance has decreased.

(b) The ratio, $(n-1)s^2/\sigma^2$, is distributed as $\chi^2$ under the assumption that the distribution of scores in the population is normal. As sample size increases, the sample distribution is more likely to approach the (nonnormal) distribution so that increasing sample size will not remedy the situation.

**7.4**

(a) The .05 and .95 critical values of $\chi^2$ are 12.592 and 1.635, and $s^2 = 178.667$. Therefore the .90 confidence limits on $\sigma^2$ are (See Equation 7.5) 85.13 and 655.66. Taking square roots, and the .90 limits on $\sigma$ are 9.33 and 25.61.

(b) To test whether $\sigma = 10$ against the alternative, $\sigma > 10$, we can calculate $\chi^2 = (6)(178.667)/10^2 = 10.72$. This is less than 12.592 ($\chi_{6,.05}^2$) so we cannot reject the null hypothesis. We could have reached the same conclusion based on the .90 confidence limits. Our decision rule is: reject if $(6)(178.667)/10^2 > 12.592$; this is algebraically equivalent to: reject if $(6)(178.667)/12.5692 > 100$. But the left hand side of this inequality was 85.13 [see part (a)], clearly less than 100. So we know from the bounds of the .90 confidence interval that the null hypothesis cannot be rejected in favor of the one-tailed alternative at the .05 level.

**7.5**

(a) Appendix Table C.5 is entered with $df_1 = 4$ and $df_2 = 10$ because the boys' variance is in the numerator. The required probabilities are (i) .10 and (ii) .975.

(b) We assume the samples were drawn from two independently and normally distributed populations with the same variances.

**7.6**

(a) Assuming that the researcher has a two-tailed alternative, she wishes to find the critical values exceeded with probability equal to .025 and .975. The .025 critical value when there are 8 and 12 $df$ is 3.512. The .975 value is $F_{8,12,.975} = 1/F_{12,8,.025} = 1/4.20 = .238$. Alternatively, most statistical packages or web sites will provide these results. In any event, the decision rule is: reject if $F > 3.51$ or $F < .238$ where $F = s_B^2 / s_G^2$.

(b) Because the alternative is now one-tailed, we reject if $s_B^2 / s_G^2 > F_{8,12,.05}$ where $F_{8,12,.05} = 2.849$.

**7.7** Let $F = s_2^2 / s_1^2 = 30/8 = 3.75$. Reject if $F > F_{10,20,.025}$ or $F < F_{10,20,.975}$. $F_{10,20,.025} = 2.77$ and $F_{10,20,.975} = 1/F_{20,10,.025} = .29$. The variances do differ significantly because $3.75 > 2.77$.

**7.8**

(a) Because $s_{\bar{Y}}^2 = s^2/n$, the estimate of the population variance is $ns_{\bar{Y}}^2$ or, in this case, $5 \times 84 = 420$. (ii) The estimate is based only on 4 observations, the four group means. Therefore there are 3 $df$

(b) $F = 420/384 = 1.09$. There are 3 and 14 $df$ so the decision rule is reject if $F > 4.242$ or $F < .070$. We cannot reject the null hypothesis.

This problem is a model for the one-factor ANOVA of Chapter 8. The major differences are that (1) the $F$ test there is against a one-tailed alternative (that the variance estimate based on means is greater than that based on the individual scores), and (2) both estimates are based on the same data set.

**7.9**

(a) The ratio of variances is $F = 292.718/56.580 = 5.173$. We also require $F_{14,12,.975}$ and $F_{14,12,.025}$; these values are .328 and 3.206. Substituting values into Equation 7.10, the confidence interval limits are $(.328)(5.173)$ and $(3.206)(5.173) = 1.70$ and $16.58$.

(b) The decision rule is: reject $H_0$ if $F > F_{12,14,.025}$ or $F < F_{12,14,.975}$. $F_{12,14,.975} = 3.05$ and therefore $H_0$ is rejected; the variance of the girls' multiplication accuracy scores is significantly greater than that for the boys. The decision rule can be framed in terms of the confidence interval limits. Rejecting $H_0$ if $F > F_{12,14,.025}$ can be shown to be algebraically equivalent to rejecting if the upper bound is less than 1. Similarly, rejecting if $F < F_{12,14,.975}$ is equivalent to rejecting $H_0$ if the lower bound is greater than 1.

(c) A stem-and-leaf plot reveals two outliers in the boys' data: 37.5 and 41.429. When these are deleted, the variance shrinks to 22.791, less than one tenth of the original variance. This illustrates that the effect of outliers in small data sets can be very large.

(d) The data are clearly not normally distributed.

**7.10** To perform the Brown-Forsythe test, we transform the fourth grade scores by letting $X$ = the absolute value of each score from the median of its gender group. For girls, the transformation is $X$ = ABS(multacc - 92.857); for boys, the transformation requires subtracting 93.333 from each multacc value. The values needed to carry out a $t$ test of the means of $X$ are:

|          | Girls  | Boys    |
| -------- | ------ | ------- |
| $n$      | 15     | 13      |
| Mean     | 5.792  | 9.350   |
| Variance | 22.775 | 225.123 |

The pooled-variance $t$ on 26 $df$ is .871 and $p$ = .39; the separate-variance $t$ on 14 $df$ is .820 and $p$ = .43. Unlike the $F$ test of the variances, this test provides no grounds for concluding that there is a difference in the spread of scores in the two populations. Given tha, the relative magnitudes of the variances reversed when we deleted two outliers from the male data (see the answer to Exercise 7.11), the present result is not surprising because the Brown-Forsythe test is less sensitive to outliers and to nonnormality.

**7.11** The relevant statistics are:

| Grade | Mean  | SD    |
| ----- | ----- | ----- |
| 3     | 3.947 | 1.350 |
| 4     | 3.662 | 1.336 |
| 5     | 2.946 | 1.258 |
| 6     | 1.841 | .499  |
| 7     | 1.763 | .548  |
| 8     | 1.842 | .710  |

(a) Plots of the means as a function of grade show a steep decline from third to sixth grade, and then a leveling off at approximately 1.8 seconds. The standard deviations also decrease from the third to the sixth grade with the sharpest drop occurring between grades 5 and 6. Surprisingly, variability increases from the sixth to the eight grade. This could be investigated by plotting the distributions. Outlying scores may be responsible. It is also possible that the upturn is not statistically significant.

(b) $H_0: \sigma_1^2 = \sigma_2^2; H_1: \sigma_1^2 \neq \sigma_2^2$. $F = .505/.249 = 2.028$. Because $F_{25,19,.025} = 2.44$, we cannot reject $H_0$.

## Chapter 8

### 8.1

(a) The variances will be multiplied by $100^2$.

(b) The $F$ ratio will not change because both numerator and denominator increase by the same factor.

(c) The variance is increased by the square of the constant.

(d) Adding a constant to all scores will not change the mean squares or the $F$ ratios.

(e) Because the spread of the group means changes, the $MS_A$ changes. However, adding the same constant to all scores in a group will not change the within-group variance and therefore $MS_{S/A}$ is unaffected.

### 8.2

(a) When $n = 10$, $MS_A = 10 \times$ (variance of the means). We enter the group means( 3, 4.8, and 7) into any calculator capable of providing the variance, and multiply by 10 to obtain $MS_A = 40.133$. The $MS_{S/A}$ is the average of the group variances; therefore, it equals 4.333 and the $F = 9.26$.

(b) Adding the totals and dividing by $N (= 24)$, $\overline{Y}_{..} = 6.167$. Then, $MS_A = [(6)(5 - 6.167)^2 + (8)(6 - 6.167)^2 + (10)(7 - 6.167)^2]/2 = 7.667$. $MS_{S/A}$ is calculated as the pooled group variances: $MS_{S/A} = [(5)(3.2) + (7)(4.1) + (9)(5.7)]/21 = 4.571$. $F = 1.68$.

### 8.3

(a)

| SV | SS | df | MS | F | p |
|----|-----|----|------|------|-----|
| A | 44.1 | 1 | 44.1 | .832 | .39 |
| S/A | 424.0 | 8 | 53.0 | | |

(b) $t = (27.8 - 23.6)/4.604 = .912$. Squaring $t$, $.912^2 = .832 = F$.

Note: Let $D = \overline{Y}_2 - \overline{Y}_1$. Then, in general,

$$t^2 = \frac{D^2}{s_{pooled}^2 (1/n_1 + 1/n_2)} = \frac{D^2/(1/n_1 + 1/n_2)}{s_{pooled}^2} = \frac{MS_A}{MS_{S/A}}$$

**8.4**

(a) The histograms and normality plots suggest that the three groups of scores do not come from normally distributed populations. However, the hypothesis of normality cannot be rejected; Kolmogorov-Smirnov and Shapiro-Wilk tests yield $p \geq .2$ for all three groups. The standard deviations are quite similar, ranging from 21.271 to 23.018. Consistent with these values, the $H$-spreads range from 29 to 35. However, Levene's test of homogeneity (compares the average absolute deviations from the group means) yields a $p$ value of .034. This may reflect the one outlier, present in the $A_1$ group. The Brown-Forsythe test (based on absolute deviations from group medians) does not yield a significant result; $p = .08$. In summary, the evidence is not strong enough to invalidate the subsequent ANOVA, particularly since the group sizes are equal.

(b) $F = 1532.067/489.216 = 3.132$, $p = .054$. The rank sums for the three groups are 257, 338.5, and 439.5. Substituting into Equation 8.21, the Kruskal-Wallis $H = 6.46$, distributed on 2 $df$; $p = .039$. There are two possible reasons for the discrepancy in $p$ values. The Kruskal-Wallis test may be more powerful because of the departure from normality. However, it is also possible that this test has led us to a Type 1 error; the significant result may reflect a difference in the shapes of the distributions (and the histograms do seem somewhat different), rather than a difference in location. The moral is that a test should be decided upon a priori on the basis of knowledge of the dependent variable's distribution. In the present case, we have no reason to depart from ANOVA, and we would fail to reject the null hypothesis.

**8.5**

(a) The treatment effects are $\hat{\alpha}_j = \overline{Y}_{.j} - \overline{Y}_{..} =$ -9.333   -1.400   10.733. The mean is zero as it should be. The average residuals are $\hat{\varepsilon}_{ij} = Y_{ij} - \overline{Y}_{.j}$ ; these also have a mean of zero.

(b) $15\sum_j \hat{\alpha}_j^2 = 3064.133 = SS_A$ and $\sum_j \sum_i \hat{\varepsilon}_{ij}^2 = 20547.067 = SS_{S/A}$.

**8.6**

(a) (i) Table 1: $F = 16$, $\hat{\theta}_A^2 = (MS_A - MS_{S/A})/n = (80-5)/10 = 7.5$; $\hat{\sigma}_A^2 = [(a-1)/a]\hat{\theta}_A^2 = (2/3)(7.5) = 5$; $\hat{\omega}_A^2 = \hat{\sigma}_A^2 / (\hat{\sigma}_e^2 + \hat{\sigma}_A^2) = 5/(5+5) = .5$. Table 2: $F = 8.5$; all other values are unchanged.

(ii) Increasing the sample size will generally change the estimate of $\omega^2$ because of the error component in the added scores. However, as these results suggest, the major change will be in the $F$ ratio which will increase because the contribution of the treatments ($\theta^2$) is multiplied by a larger $n$.

(b) $\eta^2 = SS_A / SS_{total}$. Table 1: $\eta^2 = 160/(160 + 135) = .54$; Table 2: $\eta^2 = 85/(85 + 60) = .59$. With $\omega^2$ fixed, increasing $n$ results in a *decrease* in $\eta^2$.

(c) (i) If $F = 1$, $MS_{S/A} = MS_A$, and therefore $\hat{\omega}_A^2 = 0$.

(ii) $\eta^2 = (a - 1)MS_A / [(a - 1)MS_A + a(n - 1)MS_{S/A}]$. If $F = 1$, this becomes $(a - 1) / (an - 1)$.

(d) $F$ reflects the sample size as well as treatment effects and error variance. As a result, reliance on $F$ (or the associated $p$ value) may cause very small effects to appear important if $n$ is very large. In Chapter 6 (Table 6.2), we saw a case in which the data set with the smaller standardized effect size resulted in the larger value of $t$, and in the present example we saw that $\omega^2$ (or Cohen's $f$) might be invariant while $F$ varied with $n$. As we also saw in the present example, $\eta^2$ is also affected by $n$.

**8.7**

(a) $p = .057$. The effects of $A$ are not significant.

(b) $\hat{\sigma}_A^2 = \left(\dfrac{a-1}{a}\right)\left(\dfrac{MS_A - MS_{S/A}}{n}\right) = 4.4$.  Cohen's $f = \sqrt{\hat{\sigma}_A^2 / \hat{\sigma}_e^2} = \sqrt{4.4 / 30} = .383$.

(c) Power = .41.

(d) Using GPOWER, we find $N = 159$, or 53 in each group. Note that even when effects are medium in size ($f = .25$), large $n$s are required to achieve even moderate power.

**8.8** $\sigma_A^2 = \sum_j (\mu_j. - \mu..)^2 / a$; $\mu_j. = 10$, 14, 18. Therefore, $\sigma_A^2 = 10.667$ and $f = \sqrt{10.667 / 30} = .596$. We require $N = 33$, or $n = 11$.

**8.9**

(a) The test of the means of the absolute deviations from the median yields $F = .472$, $p = .627$. Consistent with various measures of spread and data plots, there is no evidence of heterogeneity of variance.

(b) Normal probability plots for each text indicate that deviations from normality are small, mostly for the highest scores. Consistent with this impression, Neither the Kolmogorov-Smirnov or Shapiro-Wilk test (available in the SPSS Explore module) yield a significant result for any of the three groups; the lowest $p$ value for the Kolmogorov-Smirnov test is .181 for the HE text, probably because the distribution is close to the upper limit of 100.

(c) The rank sums for the three groups are 333, 341, and 502, and $H = 5.922$ which is distributed approximately as $\chi^2$ on 2 $df$ and has $p = .052$. Although slightly less than the $p$ value from the $F$ test, the result is again not significant.

**8.10**

(a) With respect to location, both means and medians increase from Sayhlth = 1 to Sayhlth = 4 with the most marked increase in depression scores occurring between the good (3) and fair (4) categories. Note that the lower bound of the .95

CI for Sayhlth = 4 is higher than the higher bound for the Sayhlth = 3 category. A test of the means with $df = 3$ and 323 yields $F = 16.85$, a very significant result.

The distributions are skewed and tests of normality indicate the departure from normality for each group except the Sayhlth = 4 group. The lack of significance for that group most likely reflects the small sample of 15; the normal probability plot has a clear departure from a straight line.

Standard deviations and $H$-spreads generally increase as ratings increase (i.e., as individuals rate their health as worse. All 4 tests of homogeneity provided by SPSS's Explore module yield $p < .01$.

(b) The best fitting straight line for the spread vs. level plot has a slope of .787. The usual strategy is to raise each score to a power equal to 1 - slope. This is .213. We rounded this and raised each score to .2.

(c) The test of the means still yields a very significant $F$, 14.102. The transformed scores still depart significantly from normality. However, the heterogeneity of variance has been greatly reduced; the Brown-Forsythe test has an associated $p$ value of .325, far from significant.

(d) This time the test of the means yields $F = 16.219$. The distributions of the transformed scores no longer departs significantly from the normal as attested to by nonsignificant Kolmogorov-Smirnov tests for each Sayhlth category. The variances are very similar; the Brown-Forsythe $p > .8$. With respect to meeting the assumptions underlying the ANOVA, log(Beck_D + 1) does a good job.

(e) Since the transformations had little effect on the $F$ test of means, why bother? Here are just a few reasons. First, confidence intervals and the proper interpretation of standardized effect size measures rest on the assumption of homogeneity of variance; confidence intervals are also affected by skew. Second, the log(Beck_D + 1) transformation may reduce heterogeneity in cases where the effects of variables are not as clear on the original scale. Third, if information is available about the parameters of the sampled population, an individual's $z$ score can be calculated and inferences drawn about that person's score relative to the population.

**8.11**

(a) The ANOVA for Sex = 0 (men)

| SV | SS | df | MS | F | p |
|---|---|---|---|---|---|
| Sayhlth | 806.380 | 3 | 268.793 | 16.268 | .000 |
| S/Sayhlth | 2577.624 | 156 | 16.523 | | |

The ANOVA for Sex = 1 (women):

| SV | SS | df | MS | F | p |
|---|---|---|---|---|---|
| Sayhlth | 727.784 | 3 | 242.595 | 7.579 | .000 |
| S/Sayhlth | 5217.474 | 163 | 32.009 | | |

(b) Using Equation 8.19, the estimate of the populations variance for the men is $\hat{\sigma}_A^2 = [(a-1)/N](MS_A - MS_{S/A}) = (3/160)(268.793 - 16.523) = 4.730$. Therefore, $\hat{f} = \sqrt{\hat{\sigma}_A^2 / \hat{\sigma}_e^2} = \sqrt{4.730/16.523} = .54$. Similar calculations for the female data yield $\hat{f} = \sqrt{\hat{\sigma}_A^2 / \hat{\sigma}_e^2} = \sqrt{3.783/32.009} = ..34$. The estimated effect of the Sayhlth variable for men is greater than that for women. However, according to the guidelines suggested by Cohen, the effect for men is large, and that for women nearly so.

**8.12**

(a) Means and medians are lower in the full employment group (Employ = 1) than in the other two groups. Normal probability plots show marked deviations from a straight line for all three groups and, consistent with this, the Kolmogorov-Smirnov and Shapiro-Wilk tests yield $p < .01$ for all three employment groups. As usual with the Beck_D scores, the distributions are quite skewed with the majority of scores in the 0 - 6 range but with a straggling tail reaching into the mid 20's. Heterogeneity of variance is also marked, largely because the variance is smaller in the employ = 1 group than in the other two. This apparent heterogeneity of variance is significant by the Brown-Forsythe test ($p = .039$) and by Levene's test ($p = .003$). Note that the smallest variance is heavily weighted in the ANOVA because there are about 4 times more subjects than in the other two groups.

(b) Although the means and medians still show a lower value in the employ = 1 group, the situation with the log(Beck_D + 1) transformation is different in other respects. Box plots and histograms reveal considerably more symmetric distributions than in the original scale and the Kolmogorov-Smirnov test no longer yields a significant result in any group. Although box plots show a larger $H$-spread for the employ = 2 category than for the other two, and that group has a larger variance than the others, the Brown-Forsythe $p$ value is .095.

(c) The ANOVAs are:

|  | SV | SS | df | MS | F | p |
|---|---|---|---|---|---|---|
| Beck_D | Employ | 173.357 | 2 | 86.679 | 3.00 | .051 |
|  | Error | 9333.241 | 323 | 28.895 |  |  |
|  | Total | 9506.598 | 325 |  |  |  |
| D_Log | Employ | 2.083 | 2 | 1.042 | 1.61 | .202 |
|  | Error | 209.053 | 323 | .647 |  |  |
|  | Total | 211.137 | 325 |  |  |  |

Although the trend for fully employed individuals to be less depressed is still apparent in the plots and statistics for the transformed scores, the variance of the means is on longer significant. The (marginal) significance of the result based on the original Beck_D scores is attributable to the heavy weighting of the smaller variance in the group of scores having the smallest variance. Reducing the heterogeneity of variance results in a higher $p$ value, and a nonsignificant result.

(d) For the Beck_D data, $\hat{f} = .11$ (see the answer to Exercise 8.11 for an example of the calculations); for the transformed data, $\hat{f} = .06$. Apparently, if employment status has an effect on depression scores in the sampled population, it is a small one.

## Chapter 9

**9.1**

(a) $H_0: (1/2)(\mu_{F1} + \mu_{F2}) - \mu_C \leq 0;$ $H_1: (1/2)(\mu_{F1} + \mu_{F2}) - \mu_C > 0.$ To test $H_0$, calculate

$$t = \frac{\hat{\psi}}{\sqrt{MS_{S/A}\sum_j w_j^2 / n}} = \frac{(1/2)(14.6 + 14.9) - 13.8}{\sqrt{(4)(.5^2 + .5^2 + 1^2)/20}} = 1.73$$

The error mean square is based on 5 x 20 scores, or 95 $df$; therefore, the $p$ value is .043. Reject $H_0$.

Note that the group variances are quite similar; therefore, we averaged them to obtain the error mean square, $MS_{S/A}$. Also, note that doubling the weights to yield integers (1, 1, and -2), leaves $t$ unchanged.

(b) $H_0: \mu_C - (1/2)(\mu_{I1} + \mu_{I2}) = 0;$ $H_1: (1/2)(\mu_{I1} + \mu_{I2}) - \mu_C \neq 0.$ Proceeding as in part (a), $t = 2.05/.548 = 3.74$, which is clearly significant.

(c) $H_0: (1/2)(\mu_{F1} + \mu_{F2}) - (1/2)(\mu_{I1} + \mu_{I2}) = 0;$

$H_1: (1/2)(\mu_{F1} + \mu_{F2}) - (1/2)(\mu_{I1} + \mu_{I2}) \neq 0.$

$t = 3/.447 = 6.71.$ The null hypothesis can be rejected.

**9.2**

(a) The form of the confidence interval is $\hat{\psi} \pm t_{95, FWE}s_{\hat{\psi}}$. For $K = 3$, $FWE = .10$ (two-tailed), and 95 $df$, the critical value of $t$ is 2.159. Note that this is the value exceeded by .0167 of the $t$ distribution. We arrive at this because, for the .90 confidence interval, we need the two-tailed value of $t$. Therefore, we require $(1/2)(FWE/3)$, or .0167. The standard error is $s_{\hat{\psi}} = \sqrt{MS_{S/A}\sum_j w_j^2 / n}$.

Therefore, the three confidence intervals are for parts (a), (b), and (c) of Exercise 9.1, respectively:

(i) $.95 \pm (2.159)(.548) = -.233, 2.133;$
(ii) $2.05 \pm (2.159)(.548) = .867, 3.233;$
(iii) $3 \pm (2.159)(.447) = 2.035, 3.965.$

Both null hypotheses, (ii) and (iii), can be rejected because $\psi_{hyp} = 0$ lies outside the limits.

(b) In the Scheffé method, the confidence limits are $\hat{\psi} \pm s_{\hat{\psi}}\sqrt{df_1 \cdot F_{FWE, df_1, df_2}}$. The critical value is the $F$ required for significance at the

.10 level with 4 and 95 *df*; this is $F_{.10, 4, 95} = 2.005$. Therefore, the three confidence limits are

(i) $.95 \pm (.548)(2.832) = -.602, 2.502$
(ii) $2.05 \pm (.548)(2.832) = .498, 3.602$
(iii) $3 \pm (.447)(2.832) = 1.734, 4.266$

Both of the last two contrasts are significant with *FWE* = .10; again, the respective confidence intervals do not contain $\psi_{hyp} = 0$.

**9.3**
(a) $SS_A = (10)[(24 - 18)^2 + (16 - 18)^2 + (14 - 18)^2] = 560$.
(b) (i) $SS_{\hat{\psi}_1} = (24 - 16)^2 / (2/10) = 320$.
(ii) $SS_{\hat{\psi}_2} = (20 - 14)^2 / (1.5/10) = 240$.
(iii) $SS_{\hat{\psi}_3} = (24 - 14)^2 / (2/10) = 500$.
Because the contrasts are orthogonal, $SS_{\hat{\psi}_2} + SS_{\hat{\psi}_1} = SS_A$.

(c) $SS_{\hat{\psi}_2}$ is unchanged because the contrast is orthogonal to the first contrast. However, $SS_{\hat{\psi}_3}$ is changed because it is not orthogonal to the first contrast. One way to think about this is in terms of a pie in which $\psi_1$ and $\psi_2$ represent nonoverlapping slices. Removing one doesn't change the size of the other. However, $\psi_3$ does overlap with the other two so that removing either of the other two affects the size of $\psi_3$.

**9.4**
(a) The grand mean is now $[(8)(24) + (10)(16) + (12)(14)]/30 = 17.333$. Then $SS_A = (8)(6.667)^2 + (10)(-1.333)^2 + (12)(-3.333)^2 = 506.667$.
$SS_{\hat{\psi}_1} = (8^2)/[1^2/8 \ + (-1)^2/10] = 284.444$.
$SS_{\hat{\psi}_2} = 6^2/[(1/2)^2/8 \ + (1/2)^2/10 \ + (-1)^2/12] = 257.910$.
$SS_A$ does not equal the sum of the two contrast sums of squares because the contrasts are no longer orthogonal when the *n*s are unequal.

(b) Define $\hat{\psi}_2 = [n_1/(n_1 + n_2)](\bar{Y}_{\cdot 1}) + n_2/(n_1 + n_2)](\bar{Y}_{\cdot 2}) - \bar{Y}_{\cdot 3}$ or, more simply, $\hat{\psi}_2 = n_1 \bar{Y}_{\cdot 1} + n_2(\bar{Y}_{\cdot 2}) - (n_1 + n_2)\bar{Y}_{\cdot 3}$. Then
$$SS_{\hat{\psi}_2} = \frac{[(8)(24) + (10)(16) - (18)(14)]^2}{8^2/8 \ + 10^2/10 \ + (-18)^2/12} = 222.222. \text{ Now } SS_{\hat{\psi}_1} + SS_{\hat{\psi}_2} = SS_A.$$

**9.5**
(a) $\hat{\psi}_s = \hat{\psi}/s_{pooled}$. $\hat{\psi} = \sum_j w_j \bar{Y}_{\cdot j} = 6$ and $s_{pooled} = \sqrt{MS_{error}} = 30$.
Therefore, $\hat{\psi}_s = .2$.

(b) $\hat{\psi} = (8/18)(24) + (10/18)(16) - 14 = 5.556$. $\hat{\psi}_S = \hat{\psi} / 30 = .185$. Note that the effect size should be calculated on the original scale; i.e., do not multiply by 18 to create integer weights.

**9.6** (a) $t = (9-2) / \sqrt{4(2/5)} = 5.53$. $FWE = .05$ (two-tailed), $df = 20$, and $K = 4$. Therefore, the critical values are 2.74 for the Dunn-Bonferroni method and 2.70 for the Dunnett method. In both cases, reject the null hypothesis.

(b) Use Tukey's $HSD$ method. Reject $H_0$ if 5.53 [the $t$ calculated in part (a)] $> q / \sqrt{2}$. The critical value of $q$ when $a = 5$, $df = 20$, and $FWE = .05$ is 4.23. Therefore, $q / \sqrt{2} = 2.99$. Again, the comparison of the $DJ$-$P$ and $C$ means is significant.

(c) The ordering of the critical values – 2.70, 2.74, and 2.99 for the Dunnett, D-B, and Tukey tests, respectively – indicate relative power with the Dunnett most powerful for the compariosn considered, and the Tukey, least powerful. A price (in power) is paid for the ability to test a larger family of comparisons.

(d) If $C$ is the critical statistic (2.70, 2.74, and 2.99), each of the three confidence intervals has the form $(\bar{Y}_{\cdot 1} - \bar{Y}_{\cdot 2}) \pm C\sqrt{MS_{\text{error}}(2/n)}$. Then the confidence intervals are

| Method | Lower Limit | Upper Limit |
|---|---|---|
| Dunnett | 3.58 | 10.42 |
| D-B | 3.53 | 10.47 |
| Tukey | 3.22 | 10.78 |

The order of the interval widths corresponds to the relative power of the tests; the narrower the interval, the more powerful the test.

**9.7**

(a) (i) Let $\psi = \mu_B - (1/2)(\mu_A + \mu_C)$. Then $H_0$: $\psi = 0$; $H_1$: $\psi > 0$.

(ii) $s_{\hat{\psi}}^2 = MS_{\text{error}} \sum_j w_j^2 / n_j$. The error mean square is the average of the three variances, and all $n$s equal 10. Therefore, $s_{\hat{\psi}}^2 = (70)(1.5/10) = 10.5$,

(iii) $t = \hat{\psi} / s_{\hat{\psi}} = 6.5/3.24 = 2.01$.

(b) (i) The standard $t$ test, on 27 $df$, is appropriate if this test is the only one and has been planned. For $\alpha = .05$ and a one-tailed alternative, reject $H_0$ if $t > 1.703$; therefore reject $H_0$.

(ii) The Scheffé method is appropriate here; $t$ is compared with $S = \sqrt{df_1 \cdot F_{.05,df_1,df_2}}$, where $df_1$ and $df_2$ refer to the numerator and denominator $df$. Substituting values, $S = \sqrt{(2)(3.35)} = 2.59$. We cannot reject $H_0$.

Note: we treat this as a test against a two-tailed alternative. If we are "data-snooping," we are considering all possible differences in both directions that might be suggested by the data.

## 9.8

(a) $K = 5$ and $df = 45$; therefore the critical value of $t$ is $t_{.01,45} \approx \pm 2.70$. We have $\hat{\psi} = \overline{Y}_{.5} - (1/4)(\overline{Y}_{.1} + \overline{Y}_{.2} + \overline{Y}_{.3} + \overline{Y}_{.4}) = 1.575$. Then, $t = \hat{\psi} / \sqrt{MS_{S/A}\sum_j w_j^2 / n} = 1.575 / \sqrt{(4)(5/4)/10} = 2.227$. We cannot reject $H_0$.

(b) The Scheffé criterion is $S = \pm\sqrt{df_1 \cdot F_{.05,df_1,df}}$, where $df_1 = 4$ and $df_2 = 45$. Therefore, $S = \pm\sqrt{(4)(2.58)} = \pm 3.21$; we cannot reject $H_0$.

(c) The Dunn-Bonferroni CI $= 1.575 \pm (2.70)\sqrt{(4)(5/4)/(10)} = -.33$, $3.48$. The Scheffé CI $= 1.575 \pm (3.212)\sqrt{(4)(5/4)/(10)} = -.70, 3.85$. The D-B interval is narrower because the family of tests is smaller. Note that both intervals include zero, indicating that the null hypothesis cannot be rejected.

(d) Using Tukey's $HSD$ method, $q_{.05,5,45} = 4.03$. Therefore, the critical $t$ value is $\pm 4.03 / \sqrt{2} = \pm 2.85$. We calculate $t = (\overline{Y}_{.2} - \overline{Y}_{.1}) / \sqrt{(4)(2)/10} = 1.006$. The mull hypothesis cannot be rejected.

(e) This is the usual $t$ test; the critical value is $t_{.05,45} = \pm 2.015$. We have $\hat{\psi} = (1/2)(\overline{Y}_{.1} + \overline{Y}_{.2}) - (1/3)(\overline{Y}_{.3} + \overline{Y}_{.4} + \overline{Y}_{.5}) = -.083$, and $s_{\hat{\psi}} = \sqrt{MS_{S/A}\sum_j w_j^2 / n} = \sqrt{(4)(5/6)/10} = .577$. Therefore, $t = .083/.577 = .144$, which is clearly not significant.

## 9.9

(a) The critical value of the Dunnett statistic is $d_{.05,30} = 2.62$. Reject $H_0$ if the absolute value of $t = (\overline{Y}_{.j} - \overline{Y}_{.C}) / \sqrt{MS_{error}(2/n)} > 2.62$, or $|\overline{Y}_{.j} - \overline{Y}_{.C}| > (2.62)\sqrt{MS_{error}(2/n)}$, or $|\overline{Y}_{.j} - \overline{Y}_{.C}| > (2.62)(2.390)$. Therefore, the critical distance is 6.26, and all comparisons, except $A_4$ versus $A_5$, are significant.

(b) The null hypothesis is $H_0: (\mu_1 - \mu_2) - (\mu_3 - \mu_4) = 0$. Therefore, $\hat{\psi} = (29) - (16) = 13$, $s_{\hat{\psi}} = \sqrt{(20)(4/7)} = 3.381$, and $t = 13/3.381 = 3.85$, which leads us to reject $H_0$.

(c) As in part (a), we calculate the critical distance by multiplying the critical value of $t$ by $s_{\hat{\psi}}$. $q_{.05,5,30} = 4.10$; therefore reject $H_0$ if $|t| > 4.10/\sqrt{2} = 2.90$, or if the absolute difference between two means is greater than $(2.90)\sqrt{MS_{error}(2/n)} = 6.93$.

## 9.10

(a) The difference between the means is 3.229 and $s_{\hat{\psi}} = \sqrt{MS_{error}(1/n_1 + 1/n_2)} = \sqrt{(17.452)(1/19 + 1/33)} = 1.203$. Therefore, $t = 2.68$.

(i) To reject $H_0$ using the Tukey-Kramer method, find $q_{.05,124}$ when $a = 4$. This is approximately 3.69. Then the critical value of $t$ is $3.69/\sqrt{2}$, or 2.61. The difference is significant.

(ii) With $K = 6$, and assuming two-tailed tests, the critical value of $t$ is $t_{.0042,124} = 2.68$; $p = .05$.

(iii) The T-K interval is CI $= 3.23 \pm (2.61)(1.203)$ whereas the D-B interval is $CI = 3.23 \pm (2.68)(1.203)$; the T-K has a slightly narrower interval.

(b) Substituting into Equation 9.5, $t' = 3.229 / \sqrt{34.541/19 + 5.970/33} = 3.229/1.414 = 2.28$. From Equation 9.10,

$$df' = \frac{1.414^4}{\dfrac{34.541^2}{(19^2)(18)} + \dfrac{5.97^2}{(33^2)(32)}} = 21.65$$

The critical $q$ value for $a = 4$, $df = 22$, and $FWE = .05$ is approximately 3.93. Dividing by $\sqrt{2}$ yields the critical $t$ value of 2.78. When heterogeneity of variance is taken into account, the result is no longer significant. This reflects the fact that in part (a), the standard error was too small because the smaller group variance was based on a larger sample. The Games-Howell procedure is most defensible for these data.

## 9.11

(a) $F = 86.679/28.895 = 3.00$; with 2 and 323 $df$, $p = .051$.

(b) For 3 groups, $q_{.05,323}$ is approximately 3.31; therefore, the critical value of $t_{crit} = 3.31/\sqrt{2} = 2.34$. Let $\overline{D} = \overline{Y}_{.j} - \overline{Y}_{.j'}$ and $SE = s_{\overline{D}}$. Then,

| Employment Categories | $\overline{D}$ | SE | CI |
|---|---|---|---|
| 1,2 | 1.274 | .872 | -.77, 3.31 |
| 1,3 | 1.730 | .783 | -.10, 3.56 |
| 2,3 | .456 | 1.053 | -2.01, 2.92 |

The confidence limits are $\overline{D} \pm t_{crit} SE$ where $SE = \sqrt{MS_{error}(1/n_j + 1/n_{j'})}$.

(c) Because the number of comparisons, $k$, equals 3, $\alpha = .017$ (two-tailed), and with 323 $df$, $t_{crit} = 2.408$. The values of $\overline{D}$ and $SE$ are the same as in part (b); the three confidence intervals now are:

| Employment Categories | CI |
|---|---|
| 1,2 | -.83, 3.37 |
| 1,3 | -.16, 3.62 |
| 2,3 | -2.08, 2.99 |

The D-B intervals are slightly wider than the T-K.

**9.12** (a) The Brown-Forsythe $F = 3.27$; with 2 and 323 $df$, $p = .039$. The variance is smallest in the largest sample (category 1). This will lead to an underestimate of the standard error and therefore to inflation of the Type 1 error rate (positive bias) and to too narrow a confidence interval for comparisons involving that category. Standard errors based on separate variances, and revised degrees of freedom (as in the Welch $t'$ test and the Games-Howell method), are more appropriate with these data.

(b) We assume heterogeneous variances and also unequal population sizes. Because of the second assumption, we have $H_0$: $[(46/106)\mu_2 + (60/106)\mu_3] - \mu_1 = 0$. To simplify computations, define $\hat{\psi} = (46\overline{Y}_{.2} + 60\overline{Y}_{.3}) - 106\overline{Y}_{.1} = 162.422$.

$$SE = \sqrt{\sum_j (w_j^2/n_j)s_j^2} = \sqrt{[(106^2)(21.825)/220] + (46)(39.882) + (60)(46.670)}$$

$= 75.861$, and $t' = 162.422/75.861 = 2.141$. The degrees of freedom are

$$df' = \frac{SE^4}{\sum_j \dfrac{w_j^4 s_j^4}{n_j^2(n_j-1)}} = 154.845.$$ The $p$ value is .034; with $\alpha = .05$, we can reject the null hypothesis.

# Chapter 10

**10.1**

(a) $\overline{X}. = 2.5$, $\overline{Y}.. = 4.84$, and

| | | | | |
|---|---|---|---|---|
| $X_j - \overline{X}. =$ | -1.5 | -.5 | .5 | 1.5 |
| $\overline{Y}._j - \overline{Y}.. =$ | 1.65 | -.02 | -.59 | -1.04 |

Let $SP = \sum_j (X_j - \overline{X})(\overline{Y}._j - \overline{Y}..) = -4.32$ and $\sum_j (X_j - \overline{X})^2 = 5.0$. Then $b_1 = SP / \sum_j (X_j - \overline{X})^2 = -.864$.

(b) $\overline{Y}_{pre,j} = \overline{Y}.. + b_1(X_j - \overline{X}) = 6.136, 5.272, 4.408,$ and $3.544$.

(c) (i) $SS_{lin} = n\sum_j (\overline{Y}_{pre,j} - \overline{Y}..)^2 = 29.860.$

(ii) $SS_{lin} = \dfrac{\hat{\psi}^2}{\sum w_j^2 / n} = \dfrac{(-8.64)^2}{(9+1+1+9)/8} = 29.860.$

$F_{1,28} = 29.860/1.42 = 21.028.$ (iii) The best-fitting straight line has a slope significantly different from zero.

**10.2**

(a) $SS_A - SS_{lin} = SS_{nonlin} = 33.221 - 29.860 = 3.361.$ This represents the variability of the points about the best-fitting straight line. It is distributed on $a - 2$ $df$ so $MS_{nonlin} = 3.361/2 = 1.681$, and $F_{2,28} = 1.681/1.42 = 1.18; p = .32$ and the result is clearly not significant. Although we cannot accept the null hypothesis of no curvature, the results are consistent with the assumption that the population means are fit by a straight line.

(b) The potential advantage of this procedure is that the error mean square is based on more degrees of freedom than the test in Exercise 10.1. The underlying assumption is that any departure from linearity reflects only error variance. If this assumption is not valid, variance due to curvature will contribute to the error mean square, negatively biasing the $F$ ratio.

**10.3** The ANOVA table is:

| SV | df | SS | MS | F | p |
|---|---|---|---|---|---|
| A | 3 | 58.000 | 19.333 | 3.69 | .034 |
| Linear | 1 | 38.440 | 38.440 | 7.33 | .016 |
| Quadratic | 1 | 12.800 | 12.800 | 2.44 | .138 |
| Cubic | 1 | 6.760 | 6.760 | 1.29 | .273 |
| S/A | 16 | 83.827 | 5.239 | | |

The results support Smith's hypothesis of an increasing trend with increased group size. However, $n$ is small and a more powerful test might reveal the significant quadratic component that Brown predicted.

**10.4**

(a) $b_1' = \sum_j \xi_{\text{lin},j} \bar{Y}_{\cdot j} / \sum_j \xi_{\text{lin},j}^2 = 12.4/20 = .62$. The standard error is $SE = \sqrt{(5.239)/(5 \times 20)} = .229$. Dividing, $t = b_1' / SE = 2.709$, and $t^2 = 7.33$, the value of $F$ in our answer to Exercise 10.3.

(b) The confidence interval for $b_1'$ is $b_1' \pm t_{.05, df} \, SE = .62 \pm (2.12)(.229) = .134$, 1.105. However, this is the confidence interval for $b_1'$, not $b_1$. The coefficients on the original $X$ scale have been multiplied by 2 (note the column labeled $\lambda$ in Appendix Table C.6) so that the $\xi_{\text{lin}}$ coefficients have integer values. To return to the original scale, divide by 2, yielding CI = .67, .56. The slope on the original scale is $(1/2)(.62) = .31$.

**10.5**

(a) According to the theoretical model, the function relating $d'$ and time should increase with time and be $S$-shaped. This suggests linear and cubic polynomial components.

(b) The sums of squares for the polynomial components can be obtained by substitution into Equation 10.14. The ANOVA results are:

| SV | df | SS | MS | F | p |
|----|----|----|----|----|----|
| Time | 6 | 8.700 | 1.450 | 20.47 | .000 |
| Linear | 1 | 5.582 | 5.582 | 78.81 | .000 |
| Quadratic | 1 | .066 | .066 | .94 | .339 |
| Cubic | 1 | 2.727 | 2.727 | 38.50 | .000 |
| Order 4 | 1 | .000 | .000 | .00 | .982 |
| Order 5 | 1 | .253 | .253 | 3.57 | .067 |
| Order 6 | 1 | .070 | .070 | .99 | .326 |
| S/Time | 35 | 2.485 | .071 | | |

As the model predicts, only the linear and cubic components are significant.

**10.6** In order to plot a function based only on the linear and cubic components, we require values of the grand mean, the group means, and the linear and cubic coefficients (from Appendix Table C.6). Then substituting into Equations 10.16 for $b_0'$ and 10.17 for $b_1'$ and $b_3'$, we have $b_0' = 2.521, b_1' = .182$, and $b_3' = .275$. To obtain the predicted mean values of $d'$ at each time interval, $\overline{Y}_{pre,j} = b_0' + b_1'\xi_1 + b_3'\xi_3$. The observed and predicted means are:

| | | | | | | | |
|---|---|---|---|---|---|---|---|
| $\overline{Y}_{\cdot j} =$ | 1.782 | 2.322 | 2.745 | 2.403 | 2.335 | 2.680 | 3.381 |
| $\overline{Y}_{pre,j} =$ | 1.699 | 2.432 | 2.614 | 2.521 | 2.428 | 2.610 | 3.343 |

The plot of the predicted as a function of time is an estimate of the population function based only on those components that were significant in the analysis.

**10. 7**
(a) The following is a plot containing SEM bars.

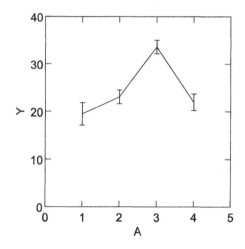

(b) The ANOVA Table is:

| SV | df | SS | MS | F | p |
|---|---|---|---|---|---|
| A | 3 | 1160.1 | 386.7 | 13.651 | .000 |
| Linear | 1 | 162.0 | 162.0 | 5.719 | .022 |
| Quadratic | 1 | 577.6 | 577.6 | 20.390 | .000 |
| Cubic | 1 | 420.5 | 420.5 | 14.844 | .000 |
| S/A | 36 | 1019.8 | 28.328 | | |

(c) The means at the $A_1$, $A_2$, and $A_3$ levels are increasing, contributing to the significant linear component. The downturn at $A_4$ is apparently responsible for the quadratic component. The inflection point at $A_2$ is the most likely reason for the cubic contribution.

**10.8**

(a) $b_3' = \sum_j \xi_{\text{cubic},j} \overline{Y}_{\cdot j} / \sum_j \xi_{\text{cubic},j}^2 = -1.45$. Multiplying each group mean by this, we obtain the the cubic components of the data: From $A_1$ to $A_4$, they are 1.45, -4.35, 4.35, and -1,45.

(b) $SS_V = 739.6$. Note that $SS_V = SS_{\text{lin}} + SS_{\text{quad}}$ as calculated in Exercise 10.7.

(c) The linear and quadratic sums of squares for $V$ are identical to those calculated in Exercise 8.10. Because the components are orthogonal, the removal of the sum of squares attributable to any one component does not affect the remaining sums of squares. The pieces of the polynomial pie don't overlap.

## Chapter 11

**11.1**

(a) The cell and marginal means are:

|  | $A_1$ | $A_2$ | $A_3$ | $A_4$ | $\overline{Y}_{\cdot\cdot k}$ |
|---|---|---|---|---|---|
| $B_1$ | 17.333 | 26.667 | 35.667 | 19.333 | 24.750 |
| $B_2$ | 28.000 | 27.000 | 20.000 | 38.333 | 28.333 |
| $\overline{Y}_{\cdot j \cdot}$ | 22.667 | 26.834 | 23.824 | 28.833 | 26.542 |

In a plot of the cell means the only clear pattern is one of interaction. The $B_1$ means increase as $A$ does until $A_3$ whereas the $B_2$ means exhibit the opposite pattern, falling until $A_3$, and then rising.

(b) The main effects are estimated by subtracting marginal means from the grand mean. For example, $\hat{\alpha}_1 = \overline{Y}_{\cdot 1} - \overline{Y}_{\cdots} = 22.667 - 26.542$. Therefore, the estimates for the $A$ and $B$ main effects are:

| $A_1$ | $A_2$ | $A_3$ | $A_4$ | $B_1$ | $B_2$ |
|---|---|---|---|---|---|
| -3.875 | .292 | 1.292 | 2.291 | -1.792 | 1.792 |

The interaction effects are obtained as $(\overline{Y}\cdot_{jk} - \overline{Y}...) - \hat{\alpha}_j - \hat{\beta}_k$ :

|       | $A_2$   | $A_3$   | $A_4$   | $A_4$   |
|-------|---------|---------|---------|---------|
| $B_1$ | -3.542  | 1.625   | 9.625   | -7.708  |
| $B_2$ | 3.542   | -1.625  | -9.625  | 7.708   |

Note that the sum of the row and column interaction entries equal zero.

(c) $SS_A = (20)[(-3.875)^2 + .292^2 + 1.292^2 + 2.291^2] = 440.423$; $SS_B = (40)(2)(1.792^2) = 256.794$; $SS_{AB} = (10)[(-3.542)^2 + ...7.708^2] = 3344.960$.

## 11.2

(a) The analysis can be readily performed on a calculator with a mean and variance key. $MS_A = bn \ x \ var(\overline{Y}\cdot_j\cdot) = 7.350$ and, because there is one $df$, $SS_A = MS_A$. Similarly, $SS_B = (b - 1) \ x \ an \ x \ var(\overline{Y}\cdot\cdot_k)$. $SS_{cells} = (ab - 1) \ x \ n \ x \ var(\overline{Y}\cdot_{jk})$. Finally, $SS_{AB} = SS_{cells} - SS_A - SS_B$. The ANOVA table is:

| SV    | df | SS      | MS     | F    | p    |
|-------|----|---------|--------|------|------|
| A     | 1  | 7.350   | 7.350  | 2.10 | .153 |
| B     | 2  | 23.333  | 11.657 | 3.33 | .043 |
| AB    | 2  | 57.600  | 28.800 | 8.23 | .001 |
| S/AB  | 54 | 189.000 | 3.500  |      |      |

(b) $SS_{A/B_3} = (6.5 - 3.4)^2/(2/10) = 48.05$; $MS_{S/A/B_3} = (5.50 + 3.75)/2 = 3.625$. Therefore, $F = 48.05/3.625 = 13.255$. On 1 and 54 $df$, $p = .001$.

(c) $MS_{B/A_2} = 10 \ x \ var(\overline{Y}\cdot_{2k}) = 2.233$; $MS_{S/B/A_2} = (1.75 + 2.25 + 3.75)/3 = 2.583$. $F < 1$ and clearly not significant.

(d) The cell variances range from 1.75 to 5.50. Unless there is clear evidence of homogeneity of variance, the safest course is to base the error term for tests of simple effects on only those cells that are involved. As we discussed in Section 9.3.2, the inclusion of variances that differ from those on which the targeted means are based can bias the $F$ test.

## 11.3

(a) Calculate the $A$, $B$, and $AB$ mean squares from the cell means. We also know that $MS_{AB}/MS_{S/AB} = 8.0$ so that we can solve for $MS_{S/AB}$. The result is

| SV | df | SS | MS | F | p |
|----|----|----|----|----|----|
| A | 1 | 4 | 4 | .5 | .485 |
| B | 2 | 32 | 16 | 2.0 | .153 |
| AB | 2 | 128 | 64 | 8.0 | .002 |
| S/AB | 30 | 240 | 8 | | |

(b) $MS_{B/A_2} = 56$, $F_{2,30} = 7.00$, $p = .003$. We assume that the six population variances are equal because we are using the error mean square from the omnibus $F$ tests.

**11.4**

(a) $Y$ will increase with both $D$ and $P$, implying that both main effects should be significant. However, the rate of increase with either will depend on the value of the other, implying a $D \times P$ interaction.

(b) $R$ will increase with both $A$ and $I$. the two sets of effects are additive so the model predicts no interaction.

(c) The significant sources of variance should be *therapy* and *therapy x socio-economic level*. Means consistent with this would be:

| | SE Level | | | |
|---|---|---|---|---|
| | Low | Middle | High | Mean |
| Psychotherapy | 10 | 12 | 14 | 12 |
| Behavior therapy | 20 | 18 | 16 | 18 |
| Mean | 15 | 15 | 15 | 15 |

**11.5** $SS_A = [(\overline{Y}_{\cdot 11} + \overline{Y}_{\cdot 12}) - (\overline{Y}_{\cdot 21} + \overline{Y}_{\cdot 22})]^2 / (4/10) = 399.361$;
$SS_B = [(\overline{Y}_{\cdot 11} + \overline{Y}_{\cdot 21}) - (\overline{Y}_{\cdot 12} + \overline{Y}_{\cdot 22})]^2 / (4/10) = 121.522$;
$SS_{AB} = [(\overline{Y}_{\cdot 11} + \overline{Y}_{\cdot 22}) - (\overline{Y}_{\cdot 12} + \overline{Y}_{\cdot 21})]^2 / (4/10) = 43.389$.

**11.6**

(a) The marginal $A$ means are 13, 10, and 7, respectively. To simplify calculations, multiply weights for $\psi_1$ by 2. Then, $SS_{\hat{\psi}_1} = [(2)(13) + (-1)(10 + 7)]^2 / [2^2/20 + (2)(-1)^2/20] = 270$. $SS_{\hat{\psi}_2} = (10 - 7)^2/(2/20) = 90$.

(b) The two contrasts are orthogonal: the sum of the crossproducts of the weights equals zero.

(c) $SS_A = (20)[(13 - 10)^2 + 0 + (7 - 10)^2] = 360$. Because the two contrasts are orthogonal and exhaust the $a - 1$ $df$, their sum equals $SS_A$.

(d) $SS_{\hat{\psi}_1 \times B} = \{[(2)(20) - (10 + 6)] - [(2)(6) - (10 + 8)]\}^2/(12/10) = 750$. $SS_{\hat{\psi}_2 \times B} = [(10 - 6) - (10 - 8)]^2/(4/10) = 10$. $SS_{AB} = 760 - (360 + 240) = 760$. The two contrasts in this part are orthogonal and exhaust the $AB$ $df$. Their sum equals $SS_{AB}$.

**11.7**

(a)

| SV | df | SS | MS | F | p |
|----|----|----|----|----|----|
| A | 2 | 12,915.56 | 6,457.78 | 82.56 | .000 |
| B | 2 | 1,202.22 | 601.11 | 7.68 | .001 |
| AB | 4 | 431.11 | 107.78 | 1.38 | .248 |
| S/AB | 81 | 6,336.00 | 78.22 | | |

(b) $\hat{\psi} = (1/2)(68.333 + 67) - 60 = 7.667$. $s_{\hat{\psi}} = \sqrt{MS_{S/AB} \sum_k w_{.k}^2 / an} = \sqrt{78.222(1.5/30)} = 1.978$. The confidence interval is $7.667 \pm (1.99)(1.978) = 3.73, 11.60$. The contrast differs significantly from zero.

(c) The contrast is $\hat{\psi} = [(1/2)(\overline{Y}_{.12} + \overline{Y}_{.13}) - \overline{Y}_{.11}] - [(1/2)(\overline{Y}_{.32} + \overline{Y}_{.33}) - \overline{Y}_{.31}]$ $= [(1/2)(56 + 52) - 44] - [(1/2)(83 + 79) - 78] = 7$, and $s_{\hat{\psi}} = \sqrt{MS_{S/AB} \sum_k w_{.k}^2 / n}$ $= \sqrt{(78.222)(3/10)} = 4.844$. $t = 7/4.844 = 1.45$; on 81 $df$, this is not significant. We cannot conclude that the contrast is different at the two age levels.

**11.8**

(a) The hypothesis suggests that both the linear and quadratic components of the rate variable should be significant. $\hat{\psi}_1 = [(-1)(60.313) + 0 + (1)(44.625)] = -15.688$ and $SS_{\text{lin(rate)}} = 15.688^2/(1/16 + 1/16) = 1968.71$. Similarly, $\hat{\psi}_2 = [(-1)(60.313) + (2)(54.813) + (-1)(44.625)] = 4.688$ and $SS_{\text{quad(rate)}} = 4.688^2/(1/16 + 4/16 + 1/16) = 58.61$. $MS_{\text{error}} = 62.152$. Therefore, the linear trend is significant ($F_{1,42} = 31.677$, $p = .000$) whereas the quadratic is not ($F < 1$).

(b) The lin (rate) $x$ text contrast is calculated as $[(-1)(66.250) + 0 + (1)(43.375)] - [(-1)(54.375) + 0 + (1)(45.875)] = -14.375$.

Squaring this and dividing by $\sum_j \sum_k w_{jk}^2 / n$, or 4/8, $SS_{\text{lin(rate)} x \text{ text}} = 413.281$.
Similarly, $SS_{\text{quad(rate)} x \text{ text}} = 78.844$. The slopes of the functions differ significantly as a function of whether the text is intact or scrambled ($F_{1,42} = 6.65$, $p = .014$). The quadratic components do not differ significantly ($F_{1,42} = 1.27$, $p = .266$.

(c) $SS_{\text{rate} x \text{ text}} = 492.125 = 413.281 + 78.844$; the two sums of squares calculated in part (b) account for the variability due to interaction.

(d) Average scores deteriorate as rate increases, but the slope is significantly more negative in the intact condition, perhaps because in the scrambled condition, performance is poor even at the slow rate. There is on evidence for quadratic curvature, in either the quadratic component of rate, or in its interaction with the type of text.

**11.9**

(a) The means are:

| Age = | 5 | 6 | 7 | 8 |
|---|---|---|---|---|
| $\overline{Y}_{..k}$ = | 14.20 | 7.25 | 6.60 | 6.80 |

The ANOVA table is:

| SV | df | SS | MS | F | p |
|---|---|---|---|---|---|
| Age | 3 | 403.72 | 134.57 | 15.38 | .000 |
| Linear | 1 | 261.06 | 261.06 | 29.84 | .000 |
| Quadratic | 1 | 127.81 | 127.81 | 14.61 | .001 |
| Cubic | 1 | 14.85 | 14.85 | 1.70 | .202 |
| Sex | 1 | 16.26 | 16.26 | 1.86 | .182 |
| Age x Sex | 3 | 31.57 | 10.52 | 1.20 | .326 |
| S/A x Sex | 32 | 280.000 | 8.75 | | |

The slope of the best fitting straight line differs significantly from zero, and there is significant (quadratic) curvature.

(b) The $SS_{\text{Age} x \text{ Sex}}$ can be partitioned into three components representing the difference between the sexes in polynomial components. For example, one component is $SS_{\text{lin(Age)} x \text{ Sex}} = \left(\sum_j \sum_k w_{jk} \overline{Y}_{\cdot jk}\right)^2 / \left(\sum_j \sum_k w_{jk}^2 / n\right)$ where $w_{1k}$ = <-3, -1, 1, 3> and $w_{2k}$ = <3, 1, -1, -3>. Age x Sex and its component terms are:

| SV | df | SS | MS | F | p |
|---|---|---|---|---|---|
| Age x Sex | 3 | 31.57 | 10.52 | 1.20 | .326 |
| lin(Age) x Sex | 1 | 10.35 | 10.35 | 1.18 | .285 |
| quad(Age) x Sex | 1 | 20.31 | 20.31 | 2.32 | .138 |
| cubic(Age) x Sex | 1 | .91 | .91 | .10 | .754 |

There are no significant differences between the sexes with respect to any of the three polynomial components.

**11.10** For the boys ($j = 1$), the linear regression coefficient (the slope of the best-fitting straight line) is $b'_{1,1} = \left( \sum_k \xi_{k1} \overline{Y}_{\cdot 1k} \right) / \left( \sum_k \xi_{k1}^2 \right) =$ [(-3)(16.3) + (-1)(7.2) + (1)(6.5) + (3)(7.4)]/20 = -1.37. Similarly, the slope of the female means is -.915. The intercepts are for the boys, $b'_{0,1} = \overline{Y}_{\cdot 1} \cdot = 9.350$, and for the girls, $b'_{0,2} = \overline{Y}_{\cdot 2} \cdot = 8.075$. Then the predicted means are $\overline{Y}_{pre,1,k} = b'_{0,1} + b'_{1,1} \xi_{\text{lin},k}$. Substituting, the predicted scores are for boys: 13.46 10.72 7.98 5.24; for girls: 10.78 8.95 7.12 5.29. The $SS_{\text{Age x Sex}} = 10.35$, the value of $SS_{\text{lin(Age) x Sex}}$, in Exercise 11.10. In summary, the sum of squares for the linear component of interaction reflects departures from parallelism of the best-fitting straight lines.

**11.11**
(a) A plot of the means for the three groups suggests both a linear and a quadratic component of the *time* source of variance. The analysis of variance confirms the linear trend and the quadratic is almost significant.

| SV | df | SS | MS | F | p |
|---|---|---|---|---|---|
| Time | 3 | 35.00 | 11.67 | 6.38 | .001 |
| linear | 1 | 27.00 | 27.00 | 14.75 | .000 |
| quadratic | 1 | 6.67 | 6.67 | 3.64 | .062 |
| cubic | 1 | 1.33 | 1.33 | .73 | .398 |
| Method | 2 | 8.13 | 4.07 | 2.22 | .119 |
| T x M | 6 | 16.40 | 2.73 | 1.49 | .201 |
| S/TM | 48 | 87.84 | 1.83 | | |

(b) (i) $H_0$: $(1/2)(\beta_{11} + \beta_{12}) - \beta_{13} = 0$.

(ii). Let $w_{jk}$ equal the products of the contrast coefficients $(1/2, 1/2,$ and $-1)$ and the linear coefficients. Multiplying by 2 to yield integers, we have $w_{jk} =$

$$Time$$

| Method | 3 | 6 | 9 | 12 |
|---|---|---|---|---|
| 1 | $(1)(-3)$ | $(1)(-1)$ | $(1)(1)$ | $(1)(3)$ |
| 2 | $(1)(-3)$ | $(1)(-1)$ | $(1)(1)$ | $(1)(3)$ |
| 3 | $(-2)(-3)$ | $(-2)(-1)$ | $(-2)(1)$ | $(-2)(3)$ |

These values are then inserted into $SS_{\hat{\psi}} = \left(\sum_j \sum_k w_{jk}\overline{Y}_{\cdot jk}\right)^2 / \left(\sum_j \sum_k w_{jk}^2 / n\right)$.

The sum of squares is 10.14 and $F_{1,48} = 10.14/1.83 = 5.54$; $p = .023$ and we reject the null hypothesis; the slope is significantly flatter in the $M_3$ condition.

An equivalent approach involves averaging the first two rows ($M_1$ and $M_2$) of cell means. This average is then contrasted with the $M_3$ means. In this case,

| | cell means | | | | coefficients | | | |
|---|---|---|---|---|---|---|---|---|
| $aver(M_1, M_2) =$ | 5.400 | 5.400 | 5.900 | 8.100 | -3.000 | -1.000 | 1.000 | 3.000 |
| $M_3 =$ | 5.200 | 5.800 | 5.400 | 5.600 | 3.000 | 1.000 | -1.000 | -3.000 |

Equation 11.10 is again applied except that the sum of the squared first row of coefficients is divided by 10 because each mean is based on 10 scores.

**11.12**

(a) The cell means (and standard deviations) are:

| | $A_1$ | $A_2$ | $A_3$ | $\overline{Y}_{\cdot\cdot k}$ |
|---|---|---|---|---|
| $B_1$ | 20.273 (.967) | 23.806 (1.599) | 30.419 (1.772) | 24.833 |
| $B_2$ | 24.448 (2.239) | 24.510 (1.945) | 23.671 (1.817) | 24.210 |
| $B_3$ | 28.069 (1.517) | 25.295 (2.084) | 22.752 (2.328) | 25.372 |
| $\overline{Y}_{\cdot j\cdot}$ | 24.263 | 24.537 | 25.614 | 24.805 |

The $AB$ interaction is most evident; at $B_1$ means increase with increases in $A$ whereas at $B_3$ means decrease as $A$ increases. The marginal means of $A$ increase slightly as $A$ increases. The marginal $B$ means decrease and then increase as $B$ increases.

(b) Consistent with the description of the pattern of means, the $AB$ interaction is very significant. However, the effects of $A$ are also and those of $B$ are marginally so.

| SV | df | SS | MS | F | p |
|----|----|----|----|---|---|
| $A$ | 2 | 30.579 | 15.289 | 4.47 | .014 |
| $B$ | 2 | 20.308 | 10.154 | 2.97 | .057 |
| $AB$ | 4 | 645.718 | 161.430 | 47.19 | .000 |
| $S/AB$ | 81 | 277.101 | 3.421 | | |

(c) We used error terms based only on the groups involved in the comparison. For example, in comparing the marginal means of $A_1$ and $A_2$, we averaged the six variances corresponding to the cells in those conditions. This seemed the most conservative approach; however, the resulting error terms were very close to $MS_{S/AB}$. The results were:

| Comparison | Numerator Mean Square | Error Term | F | p |
|-----------|----------------------|-----------|---|---|
| $A_1$ vs $A_2$ | 1.126 | 3.155 | .357 | .553 |
| $A_1$ vs $A_3$ | 27.358 | 3.352 | 8.162 | .006 |
| $A_2$ vs $A_3$ | 17.385 | 3.756 | 4.628 | .036 |

All tests were carried out with 1 and 54 $df$. Because there were three planned tests, the Dunn-Bonferroni method requires that if $FWE = .05$, $p < .017$ for significance. Only the $A_1$ vs $A_3$ comparison meets that criterion.

**11.13**

(a) The causal means and variances for the eight cells are

| Instructions | Format = 1 (Text) Mean | Variance | Format = 2 (Web) Mean | Variance |
|-----|-----|-----|-----|-----|
| 1 ($N$) | .625 | 1.982 | 2.875 | 5.554 |
| 2 ($S$) | 1.625 | 3.411 | 2.500 | 1.714 |
| 3 ($E$) | 3.000 | 3.714 | 4.000 | 11.143 |
| 4 ($A$) | 4.500 | 5.429 | 6.500 | 11.714 |

In both format conditions, instructions influence mean performance, with means highest in the argument (*A*) condition, and tending to be lowest in the narrative (*N*) and summary (*S*) conditions. Means are consistently higher in the web than in the text condition. There may be an interaction with the largest format effects present in the *N* and *A* conditions. There appears to be considerable heterogeneity of variance, with variances highly correlated with the means. In fact, SPSS reports that Levene's *F* on 7 and 56 *df* equals 2.437, and is significant at the .03 level. Box plots and stem-and-leaf plots reveal a similar pattern for the *H*-spreads. Thus, although the combination of web learning and argument instruction leads to improved causal inferencing, it also leads to increased variability. This would be a problem if such variability resulted because some individuals did very poorly in the web/E condition. However, note that in both format conditions, the lower hinge in the box plot (essentially the 25th percentile is at or above the medians for the other instructional groups. This suggests that the advantage of the web/E condition is fairly general, though perhaps more so for some individuals than for others. Finally, normal probability plots, the Kolmogorov-Smirnov test, and skewness and kurtosis statistics all make clear that the data are not normally distributed. In particular, the text/N and text/S distributions are skewed to the right with a pileup of scores near the median.

(b) The ANOVA table is

| SV | df | SS | MS | F | p |
|---|---|---|---|---|---|
| format (*F*) | 1 | 26.266 | 26.266 | 4.705 | .034 |
| instruction (*I*) | 3 | 106.672 | 35.557 | 6.369 | .001 |
| *F x I* | 3 | 5.047 | 1.682 | .301 | .824 |
| *S/FI* | 56 | 312.625 | 5.583 | | |

(c) The new cell means and variances are

| | Format = 1 (Text) | | Format = 2 (Web) | |
|---|---|---|---|---|
| Instructions | Mean | Variance | Mean | Variance |
| 1 (*N*) | .288 | .344 | 1.169 | .469 |
| 2 (*S*) | .814 | .300 | 1.192 | .141 |
| 3 (*E*) | 1.254 | .365 | 1.452 | .328 |
| 4 (*A*) | 1.605 | .260 | 1.732 | .349 |

The pattern of means is much the same as on the original scale although there is a more marked trend toward interaction; the effect of format decreases as we move from the narrative to the explanation condition. The most evident change is that the differences among the variances are much reduced; in fact, the Levene test of homogeneity, which previously yielded a $p$ value of .03, now yields $F_{7,56} = .384$, $p = .91$. In view of this, it is of interest to redo the ANOVA, using the transformed scores:

| SV | df | SS | MS | F | p |
|---|---|---|---|---|---|
| format ($F$) | 1 | 2.512 | 2.512 | 7.86 | .007 |
| instruction($I$) | 3 | 8.059 | 2.686 | 8.41 | .000 |
| $F \times I$ | 3 | 1.387 | 0.462 | 1.45 | .239 |
| S/FI | 56 | 17.891 | 0.319 | | |

Although $p$ values are somewhat less than on the original data scale, conclusions are essentially the same.

## Chapter 12

**12.1**

(a) For the $A$ source of variance:

$$H_0 : (\mu_{111} + \mu_{112} + \mu_{121} + \mu_{122}) - (\mu_{211} + \mu_{212} + \mu_{221} + \mu_{222}) = 0$$

For the $BC$ source of variance:

$$H_0 : (\mu_{111} + \mu_{211} + \mu_{122} + \mu_{222}) - (\mu_{112} + \mu_{212} + \mu_{121} + \mu_{221}) = 0$$

For the $ABC$ source of variance:

$$H_0 : (\mu_{111} + \mu_{122} + \mu_{212} + \mu_{221}) - (\mu_{211} + \mu_{222} + \mu_{112} + \mu_{121}) = 0$$

(b) The coefficients follow from the answer to part (a). For example, $SS_A =$

$$\left( \sum_j \sum_k \sum_m w_{jkm} \overline{Y}_{\cdot jkm} \right)^2 \bigg/ \left( \sum_j \sum_k \sum_m w_{jkm}^2 / n \right) = [(12 + 14 + 8 + 13) - (17 + 16 +$$

$10 + 18)]^2/(8/10) = 145$. Similarly, substituting 1's and -1's for the $w$s, $SS_{BC} = 180$ and $SS_{ABC} = 45$.

**12.2**

(a) From the $df$ for $A$, $B$, and $C$, we know there are 24 cells. The $df_{tot} = 95$. Therefore, $n = 4$. Also, given the $MS_A$ and $F$, $MS_{S/ABC} = 20$.

(b) $\hat{\sigma}_A^2 = \left( \frac{a-1}{a} \right) \left( \frac{MS_A - MS_{S/ABC}}{bnc} \right) = (3/4)[(56.8 - 20)/24] = 1.15$;

$\hat{f}^2 = \hat{\sigma}_A^2 / \hat{\sigma}_e^2 = 1.15/20 = .0575$

(c) We used GPOWER's "Other $F$ test module." This requires numerator and denominator $df$ to be input; those values are 3 and 24 $x$ 7, or 168. Power =.58.

**12.3** Let $A$, $I$, and $X$ represent age, irrelevant information, and sex, respectively. Then the $SV$, $df$, and $EMS$ are:

| SV | df | EMS |
|----|----|-----|
| $A$* | 2 | $\sigma_e^2 + 60\theta_A^2$ |
| $I$* | 2 | $\sigma_e^2 + 60\theta_I^2$ |
| $X$ | 1 | $\sigma_e^2 + 90\theta_X^2$ |
| $AI$* | 4 | $\sigma_e^2 + 20\theta_{AI}^2$ |
| $AX$ | 2 | $\sigma_e^2 + 30\theta_{AX}^2$ |
| $IX$ | 2 | $\sigma_e^2 + 30\theta_{IX}^2$ |
| $AIX$* | 4 | $\sigma_e^2 + 10\theta_{AIX}^2$ |
| $S/AIX$ | 162 | $\sigma_e^2$ |

(b) Sources followed by an asterisk (*) are hypothesized to be significant. One way to approach this would be to generate means that have the effects hypothesized. For example, set $\mu = 500$. Then hypothesis (i) might imply means of 520 for the 3-year olds, 500 for the 5-year olds, and 480 for the 7-year olds. Building in the other hypothesized effects, we might have

| | Boys | | | | Girls | | |
|-------|-----|-----|-----|-------|-----|-----|-----|
| Age = | 3 | 5 | 7 | Age = | 3 | 5 | 7 |
| $I_1$ | 500 | 490 | 480 | $I_1$ | 510 | 495 | 480 |
| $I_2$ | 520 | 500 | 480 | $I_2$ | 520 | 500 | 480 |
| $I_3$ | 540 | 510 | 480 | $I_3$ | 530 | 505 | 480 |

**12.4**

(a) The following means are consistent with all the hypotheses:

| | LoA | | | | | HiA | | | |
|---|---|---|---|---|---|---|---|---|---|
| Time = | 30 | 45 | 60 | Mean | Time = | 30 | 45 | 60 | Mean |
| LP | 40 | 60 | 70 | 56.67 | LP | 60 | 70 | 75 | 68.33 |
| BP | 40 | 60 | 70 | 56.67 | BP | 60 | 70 | 75 | 68.33 |
| NP | 10 | 40 | 55 | 35.00 | NP | 30 | 50 | 60 | 46.67 |
| Mean | 30.00 | 53.33 | 65.00 | 49.44 | Mean | 50.00 | 63.33 | 70.00 | 61.11 |

(b) In accord with the hypotheses, the theoretical means predict (i) an ability main effect; (ii) An advantage of programmed instruction over the *NP* instruction; (iii) An interaction of ability *x* time; (iv) both linear and quadratic components of the time function; (v) a steeper linear function of time for the *NP* conditions.   Accordingly, all three main effects should be significant, as should the linear and quadrtic components of time. $H_2$ predicts more than a main effect of instruction; the contrast of the average of the two programmed instruction means against the *NP* mean should be a significant component of the instruction variability.  The time *x* ability interaction should be significant.  Finally, a contrast of the slope of the *NP* function of time with the average slope of the *LP and BP* conditions should be significant.  It should be noted that a three-factor interaction is not a necessary consequence of the hypotheses.

**12.5**

(a) $SS_{\text{cells}} = \sum_j \sum_j n_{jk} (\overline{Y}_{.jk} - \overline{Y}...)^2 = (2)(10 - 5)^2 + (4)(10 - 5)^2 +$ $(8)(2\text{-}5)^2 + (2)(2\text{-}5)^2 = 240$.

(b) $SS_A = \sum_j n_j . (\overline{Y}_{.j} - \overline{Y}...)^2 = (6)(10\text{-} 5)^2 + (10)(2\text{-}5)^2 = 240$; similarly, $SS_B = 52.267$. The variability due to *A* and *B* is greater then the variability among the four cell means, an impossible result.  If we were to calculate $SS_{AB}$ in the usual way (subtracting from $SS_{\text{cells}}$), we would have a negative sum of squares.  The problem is that the *A* and *B* effects are correlated.

(c) $\hat{\alpha}_1 = 10 - 5 = 5$;   $\hat{\alpha}_2 = 2 - 5 = -3$. Note that $\sum_j n_j . \hat{\alpha}_j = 0$.

(d) Adjusting for the effects of *A*, the cell means are now all 5.  The $SS_B$ and $SS_{AB}$ are now both zero.  This peculiar state of affairs exists because the *A*, *B*, and *AB* effects are perfectly correlated in this "data set."  Although the correlation won't be zero with real disproportionate cell frequencies, calculations such as those in parts (a) and (b) will give misleading, and often absurd, results.  Regression methods that provide more appropriate analyses are discussed in Chapter 20 and 21.

**12.6**

(a) (i) The row means are: $\overline{Y}_{.j.} = 14, 9, 6.67$; the column means are $\overline{Y}_{..k} = 7, 5.67, 17$. (ii) The grand mean is $\overline{Y}_{...} = 9.89$; therefore, $\hat{\alpha}_j = 4.11$, -.89, -3.22. (iii) The matrix, adjusted for the row effects, is now

|       | $B_1$ | $B_2$ | $B_3$ |
|-------|-------|-------|-------|
| $A_1$ | 7.89  | 3.89  | 17.89 |
| $A_2$ | 8.89  | 6.89  | 13.89 |
| $A_3$ | 4.22  | 6.22  | 19.22 |
| Means | 7.00  | 5.67  | 17.00 |

The column means are unchanged.

(b) (i) The row means are now 15.67, 9.83, and 9.17; the column means are 9.125, 6.625, and 17.875. (ii) The grand mean is the weighted (by row frequencies – 24/48,18/48, and 6/48) average of the row means, 12.67. Subtracting the row means from the grand mean, $\hat{\alpha}_j$ = 3.00, -2.83, and -3.50. Note that their weighted average equals zero. (iii) Subtracting the $\hat{\alpha}_j$ for each cell mean, then finding the weighted average of each column, we find the column means unchanged.

(c) Again being careful to take weighted averages of cell means, we have $\overline{Y}_{.j.} = 12, 10$, and 5.5; $\overline{Y}_{..k} = 5.5, 6.61$, and 15.68; and $\overline{Y}_{...} = 9.25$. Then $\hat{\alpha}_j$ = 2.75, .75, and -3.75. Adjusting cell totals as before, the column means are now 7.21, 6.32, and 17.47.

(d) $A$ and $B$ effects are orthogonal when cell frequencies are equal or proportional. Removal of the effects (and therefore the sums of squares) of one variable does not affect the effects (or sums of squares) of the other. When $n$s are disproportional, the effects are correlated and the variables do not make independent contributions.

**12.7**

(a) $\hat{f}_B = \sqrt{(df_B / N)(F_B - 1)} = \sqrt{(2 / 90)(1.97)} = .21$. $\omega_B^2 = (df_B)(F_B - 1) / [(df_B)(F_B - 1) + N] = (2(1.97)/[(2)(1.97)+90] = .04$. The effect is roughly medium in size by Cohen's guidelines.

(b) The power to detect $f$ = .25 with 2 and 81 $df$ is .54. Although 90 subjects may seem like a large number in an experiment, the power is low and if it's important to detect $B$ effects (assuming they exist), a larger sample size is needed to have reasonable power.

**12.8** The means are

|  | Time = | 15 | 30 | 45 | Means |
|---|---|---|---|---|---|
| Exptl | High | .539 | .201 | .176 | .305 |
|  | Low | .216 | .041 | .059 | .105 |
| Means |  | .378 | .121 | .118 | .205 |
| Control | High | .197 | .166 | .175 | .179 |
|  | Low | .089 | .112 | .120 | .107 |
| Means |  | .143 | .139 | .148 | .143 |

There are more false reports in the experimental condition than in the control condition; the experimental mean is .062 higher than the control. However, this seems entirely due to the 15 second viewing condition. Relatedness also appears to have an effect; there were more false recalls of items that were related to those actually studied. This is true even in the control groups although the effect is greater in the experimental groups.

(b) The ANOVA table, including trend components, is

| SV | df | SS | MS | F | p |
|---|---|---|---|---|---|
| Context ($C$) | 1 | .092 | .092 | 3.39 | .069 |
| Time ($T$) | 2 | .354 | .177 | 6.51 | .002 |
| lin($T$) | 1 | .261 | .261 | 9.59 | .003 |
| quad($T$) | 1 | .093 | .093 | 3.43 | .068 |
| Related($R$) | 1 | .445 | .445 | 16.35 | .000 |
| $C \times T$ | 2 | .357 | .178 | 6.56 | .002 |
| $C \times$ lin($T$) | 1 | .280 | .280 | 10.37 | .002 |
| $C \times$ quad($T$) | 1 | .077 | .077 | 2.85 | .095 |
| $C \times R$ | 1 | .098 | .098 | 3.59 | .061 |
| $T \times R$ | 2 | .077 | .039 | 1.42 | .247 |

| | | | | | |
|---|---|---|---|---|---|
| $C \times T \times R$ | 2 | .025 | .012 | .46 | .634 |
| Error | 84 | 2.285 | .027 | | |

Objects highly related to those which actually were studied were more often falsely recalled; this effect was quite significant ($F_{1,84} = 16.35$, $p = .000$). Although false recalls occurred significantly more often in the experimental condition (in which the confederate gave false reports), this appears to be entirely due to the shortest viewing times. This is reflected in the significant context $x$ time interaction, and particularly by its linear component.

**12.9**

(a) The means (and *SEMs*) for the Wiley-Voss SVT data are:

| | | | | | |
|---|---|---|---|---|---|
| Text | 80.00 (3.27) | 80.00 (3.78) | 70.00 (5.35) | 71.25 (4.80) | 75.31 |
| Web | 71.88 (4.32) | 69.38 (3.59) | 62.50 (2.67) | 75.62 (2.74) | 69.84 |
| Means | 75.94 | 74.69 | 66.25 | 73.44 | 72.58 |

(b) The ANOVA table is:

| *SV* | *df* | *SS* | *MS* | *F* | *p* |
|---|---|---|---|---|---|
| Format | 1 | 478.52 | 478.52 | 3.90 | .053 |
| Instructions | 3 | 904.30 | 301.43 | 2.45 | .073 |
| $F \times I$ | 3 | 538.67 | 179.56 | 1.46 | .235 |
| Error | 56 | 6878.13 | 122.82 | | |

(c) $\hat{f} = \hat{\sigma}_{\text{effect}} / \hat{\sigma}_e$ and, for example in a two-factor design,
$\hat{\sigma}_A = \sqrt{[df_A / df_A + 1)](MS_A - MS_{\text{error}}) / nb}$.

| Source | $\hat{\sigma}_{\text{effect}}$ | $\hat{f}$ |
|---|---|---|
| Format | $\sqrt{(1/2)(478.52 - 122.82)/32} = 2.358$ | .21 |
| Instructions | $\sqrt{(3/4)(301.43 - 122.82/16} = 2.894$ | .26 |
| $F \times I$ | $\sqrt{(3/4)(179.56 - 122.82/8} = 2.306$ | .21 |

(d) Although the Format $F$ is larger, and its $p$ value smaller, than that for Instructions, the estimated effect of Instructions is larger. $E(MS_{Format}) = \sigma_e^2 + 32\theta_{Format}^2$ whereas $E(MS_{Instructions}) = \sigma_e^2 + 16\theta_{Instructions}^2$. The larger coefficient of the $\theta^2$ term in the Format $F$ test contributes to its larger $F$. However, the estimated ratio of population standard deviations is not affected by the coefficients.

(e) We require $N = 128$ to detect $f = .25$ with power $= .80$ for the format effects but $N = 176$ for the instructional effects. It may seem surprising that a much larger $N$ is needed to have the same power. However, note that for fixed $N$, each of the four instructional means is based on fewer scores than each of the two format means and therefore has a larger standard error. To compensate for this, a larger sample size is needed to test Instructions.

## 12.10

(a)

| Mood | Focus = Content (1) | | | Focus = Language (2) | | |
| | Message | | | Message | | |
| | Strong (1) | Weak (2) | Means | Strong (1) | Weak (2) | Means |
|---|---|---|---|---|---|---|
| Happy (1) | 52.70 | 49.10 | 50.90 | 51.30 | 56.70 | 54.00 |
| Sad (2) | 59.50 | 46.90 | 53.20 | 51.20 | 50.10 | 50.65 |
| Means | 56.10 | 48.00 | 52.05 | 51.25 | 53.40 | 52.325 |

There may be a main effect of message; average fees are higher following strong than weak messages. However, this appears to interact with the participant's focus; the strong-message mean exceeds the weak-message mean only when subject's focus on the content of the message. There appears to be a similar interaction of mood by focus. When subjects focus on the content they tend to assess higher fees when in a sad mood but the opposite is true when they focus on the language of the message. Finally, there appears to be an interaction between mood and message; as the means below indicate, the higher assessed fees with the strong message occur only when a sad mood has been induced.

| Mood | Message | |
| | Strong (1) | Weak (2) |
|---|---|---|
| Happy (1) | 52.00 | 52.90 |
| Sad (2) | 55.35 | 48.5 |

(b) The ANOVA table:

| SV | df | SS | MS | F | p |
|---|---|---|---|---|---|
| Message (M) | 1 | 177.013 | 177.013 | 8.259 | .005 |
| Mood (m) | 1 | 5.513 | 5.513 | .257 | .614 |
| Focus (F) | 1 | 1.513 | .513 | .071 | .791 |
| M x m | 1 | 300.313 | 300.313 | 14.012 | .000 |
| M x F | 1 | 525.313 | 525.313 | 24.511 | .000 |
| m x F | 1 | 159.613 | 159.613 | 7.447 | .008 |
| M x m x F | 1 | 7.813 | 7.813 | .365 | .548 |
| S/MmF | 72 | 1543.100 | 21.432 | | |

The ANOVA is consistent with the patterns described in part (a).

**12.11** Dunnett's method is the appropriate method for comparing each of several groups with a control. The control group variance ($s^2 = 4.5^2$) is quite similar to the $MS_{error}$ and therefore can be pooled with it. We can compare the $t$ statistic with the critical value of Dunnett's $d$ statistic (Appendix Table C.8), where $t = (\overline{Y}_E - \overline{Y}_C)/\sqrt{s^2_{pooled}(1/n_E + 1/n_C)}$ , and the $E$ and $C$ subscripts refer to experimental and control groups. In the present example, the pooled variance = $[(72)(21.432) + (9)(20.25)]/81 = 21.300$, $n_E = n_C = 10$, and the standard error is $s_{diff} = 2.064$. The critical value of $d$, with $\alpha = .10$ and 81 $df$, is approximately 2.43. We could calculate 8 $t$ statistics, comparing each with $d$ but a simpler approach is to note that if $t > d_{crit}$, then $|\overline{Y}_E - \overline{Y}_C| > s_{diff}d_{crit}$ , or 5.02. The absolute differences between each mean and the control mean is:

| | Focus = Content (1) | | Focus = Language (2) | |
|---|---|---|---|---|
| | Message | | Message | |
| Mood | Strong (1) | Weak (2) | Strong (1) | Weak (2) |
| Happy (1) | 4.70 | 1.10 | 3.30 | 8.70* |
| Sad (2) | 11.50* | 1.10 | 3.20 | 2.10 |

The two differences accompanied by an asterisk are the only significant ones. In summary, means of experimental conditions differ significantly from the mean of a control only when subjects are sad and focused on the content of a strongly worded message, or when subjects are happy, and focused on the language of a weakly worded message.

## Chapter 13

**13.1**

(a, b)

| SV | df | MS | F | EMS |
|---|---|---|---|---|
| Subjects | 3 | 23.556 | | $\sigma_e^2 + 3\sigma_S^2$ |
| A | 2 | 16.583 | 14.559 | $\sigma_e^2 + 4\theta_A^2$ |
| S x A | 6 | 1.139 | | $\sigma_e^2$ |

(c) $\hat{\sigma}_A^2 = [(a-1)/a]\hat{\theta}_A^2 = [(a-1)/a][(MS_A - MS_{SxA})/n]$
$= (2/3)(3.861) = 2.574; \hat{\sigma}_S^2 = (MS_S - MS_{SxA})/a = 7.472;$ partial-$\hat{\omega}_A^2 =$
$\hat{\sigma}_A^2 / (\hat{\sigma}_A^2 + \hat{\sigma}_e^2) = 2.574/(2.574+1.139) = .69.$

(d) est $MS_{S/A} = [n(a-1)/(an-1)]MS_{SxA} + [(n-1)/(an-1)]MS_S$
$= (8/11)(1.139) + (3/11)(23.556) = 7.253.$ Note how much larger the error variance is estimated to be if we use a between-subjects design.

(e) For the repeated-measures design, assuming sphericity, calculate
$\hat{f} = \hat{\sigma}_A / \hat{\sigma}_e = \sqrt{2.574/1.139} = 1.50$ or $\hat{\lambda} = N\hat{f}^2 = (12)(2.25) = 27.$ Then, power $\approx .95.$

For the between-subjects design, $\hat{f} = \hat{\sigma}_A / \hat{\sigma}_e = \sqrt{2.574/7.253} = .596$
and $\hat{\lambda} = N\hat{f}^2 = (12)(.355) = 4.259.$ Then power = .28. The loss in power would not be as great if $N$ were larger; nevertheless, the example illustrates the difference in the two designs, and the reason for it – the difference in error variance.

**13.2**

(a)

| Subject | $d_{12}$ | $d_{13}$ | $d_{23}$ |
|---|---|---|---|
| 1 | 0.7 | 1.0 | 0.3 |
| 2 | 1.7 | 2.4 | 0.7 |
| 3 | -0.1 | 3.3 | 3.4 |
| 4 | 2.5 | 3.9 | 1.4 |
| 5 | 0.1 | 3.9 | 3.8 |
| 6 | 0.1 | 0.7 | 0.6 |

The variances are 1.099, 2.011, and 2.312, respectively.

    (b) The correlation matrix is

$$\begin{array}{c c c c} & A_1 & A_2 & A_3 \\ A_1 & 1.000 & & \\ A_2 & 0.804 & 1.000 & \\ A_3 & 0.843 & 0.817 & 1.000 \end{array}$$

The covariance of $A_j$ and $A_k$, $s_{jk}$, $= r_{jk}s_j s_k$. Substituting, the variance-covariance matrix is

$$\begin{array}{c c c c} & A_1 & A_2 & A_3 \\ A_1 & 2.943 & & \\ A_2 & 2.228 & 2.612 & \\ A_3 & 3.646 & 3.330 & 6.360 \end{array}$$

Substituting values from the preceding matrix, the variance of the difference score, $var(d_{jk})$, $= s_j^2 + s_k^2 - 2r_{jk}s_j s_k = 1.099$, 2.011, and 2.312, the values calculated in part (a).

    (c)

| SV | df | MS | F |
| --- | --- | --- | --- |
| A | 2 | 10.002 | 11.07 |
| S x A | 10 | .904 | |

$(1/2)$ $x$ the average of the three values of $var(d_{jk})$ is $(1/2)[(1.099 + 2.011 + 2.312)/3] = .904$, the error mean square in the ANOVA.

    (d) The two tests are $t_{10} = (5.2 - 4.367)/\sqrt{.904/6} = 2.146$; $.05 < p < .1$; and $t_5 = (5.2 - 4.367)/\sqrt{1.099/6} = 1.946$; $.1 < p < .2$. Because the values of $var(d_{jk})$ are not homogeneous, the $MS_{SxA}$ gives a positively biased result. The test based on $var(d_{12})$ is the correct test.

    (e) The average of the six contrasts is $\bar{\psi} = (1/2)(5.2 + 6.9) - 4.367 = 1.683$. The variance of the linear contrast can be most easily obtained by calculating a contrast score, $\hat{\psi}_i$ for each subject; then the variance of these "scores" is calculated as usual: $var(\hat{\psi}) = \sum_i (\hat{\psi}_i - \bar{\psi})^2 / n = .977$. Finally,

$t = 1.683 / \sqrt{.977 / 6} = 4.172$, which is clearly significant at the .05 level.

**13.3** The matrix does not exhibit compound symmetry because the variances are not equal, nor are the covariances. However, it does meet the sphericity definition because $var(d_{12}) = 1 + 3 - (2)(.5) = 3$; $var(d_{13}) = 1 + 5 - (2)(1.5) = 3$; and $var(d_{23}) = 3 + 5 - (2)(2.5) = 3$. The variances of difference scores are identical.

**13.4**

(a)

| SV | SS | df | MS | F | p |
|----|------|----|-------|-------|-------|
| A | 3.763 | 3 | 1.254 | 6.350 | 0.003 |
| SxA | 4.148 | 21 | 0.198 | | |

If $\varepsilon = 1$, the df are unchanged, and $p = .003$. At its lower limit, $\varepsilon = 1/(a-1)$; then $df = 1$ and 7, and $p = .04$. Therefore, $.003 < p < .04$; in either case, we conclude that the $\mu_j$ are not all equal.

(b) Using the Dunn-Bonferroni method, if the FWE = .05, $\alpha = (.05/6) = .008$ (two-tailed). We want the value of $t$ on 1 and 7 df that cuts off .004 in each tail. This is approximately 3.64. The means are 3.275 and 4.175 and the standard deviation of the difference is .804. Therefore, the .95 simultaneous confidence interval for this difference is $(4.175 - 3.275) \pm (3.64)(.804/\sqrt{8}) = -.135, 1.925$.

**13.5** Following is SPSS's output for a trend analysis:

**Tests of Within-Subjects Contrasts**

Measure: MEASURE_1

| Source | A | Type III Sum of Squares | df | Mean Square | F | Sig. |
|--------|-----------|--------|----|--------|--------|------|
| A | Linear | 2.809 | 1 | 2.809 | 10.566 | .014 |
| | Quadratic | .911 | 1 | .911 | 3.777 | .093 |
| | Cubic | 4.225E-02 | 1 | 4.225E-02 | .495 | .505 |
| Error(A) | Linear | 1.861 | 7 | .266 | | |
| | Quadratic | 1.689 | 7 | .241 | | |
| | Cubic | .598 | 7 | 8.539E-02 | | |

The coefficients for each polynomial component can be found in Appendix Table C.6. For the linear component of $A$, $\left(\sum_j \xi_j \overline{Y}_{\cdot j}\right)^2 / \left(\sum_j \xi_j^2 / n\right) = [(-3)(3.413)$
$+ (-1)(3.275) + (1)(3.638) + (3)(4.175)]^2/(20/8) = 2.807$, SPSS's answer, within rounding error. The remaining components of the $A$ sum of squares are calculated in the same way; only the coefficients change. One way to calculate the error components (other than using a program that does it for us, as we did) is $SS_{\lin(A)xS} =$

$\sum_i \left( \sum_j \xi_j Y_{ij} \right)^2 / \sum_j \xi_j^2$ - $SS_{\text{lin}(A)}$. An alternative way to carry out the $F$ test is to transform the scores.. For example, $S_1$'s linear score is $C_{1,\text{lin}} = (-3)(1.8) + (-1)(2.2) + (1)(3.2) + (3)(2.4) = 2.8$. Once this transformation is carried out for each subject, $F = t^2 = \overline{C}^2 / (s_C^2 / n)$.

**13.6**

(a) $MS_e = (19/760)(739{,}141) + (741/760)(853.157) = 850.307$.

(b) $\hat{\sigma}_S^2 = (MS_S - MS_e) / a = (208{,}305.017 - 850.307)/20 = 10{,}372.736$.

(c) $r_{11} = \hat{\sigma}_S^2 / (\hat{\sigma}_S^2 + \hat{\sigma}_e^2) = 10{,}372.736/(10{,}372.736 + 850.357) = .92$.

**13.7**

(a) $\hat{X}_{ij} = (nT_{i.} + aT_{.j} - T..) / [(n-1)(a-1)]$

$= [(4)(61) + (3)(78) - 331] / [(3)(2)] = 24.5$

(b) In the first cycle, we initially set $\hat{X}_{12} = 30.5$, the average of the remaining two scores for $S_1$; however, the initial estimate is not critical. We then proceed as in part (a) to estimate $\hat{X}_{43}$. In the next cycle, we use that estimate, together with the observed scores, to get a new estimate of $\hat{X}_{12}$, and so on. Our results are:

| | Cycles | | | |
|---|---|---|---|---|
| | 1 | 2 | 3 | 4 |
| $\hat{X}_{12}$ | 30.5000 | 24.6806 | 24.5189 | 24.5144 |
| $\hat{X}_{43}$ | 37.9167 | 38.8866 | 38.9135 | 38.9143 |

**13.8**

(a)

| SV | EMS |
|---|---|
| W | $\sigma_e^2 + 4\sigma_{WO}^2 + 20\sigma_W^2$ |
| P | $\sigma_e^2 + 10\sigma_{PO}^2 + 5\sigma_{WP}^2 + \sigma_{WPO}^2 + 50\theta_P^2$ |
| O | $\sigma_e^2 + 4\sigma_{WO}^2 + 40\sigma_O^2$ |
| WxP | $\sigma_e^2 + 5\sigma_{WP}^2 + \sigma_{WPO}^2$ |
| WxO | $\sigma_e^2 + 4\sigma_{WO}^2$ |
| PxO | $\sigma_e^2 + 10\sigma_{PO}^2 + \sigma_{WPO}^2$ |
| WxPxO | $\sigma_e^2 + \sigma_{WPO}^2$ |

(b) Applying Equation 13.17, $F_1' = \dfrac{MS_P}{MS_{PO} + MS_{WP} - MS_{WPO}} =$

$\dfrac{2610}{330 + 640 - 320} = 4.015$, The denominator $df$ are (Equation 13.18)

$$df_{error}' = \frac{(MS_{PO} + MS_{WP} - MS_{WPO})^2}{\dfrac{MS_{PO}^2}{(p-1)(o-1)} + \dfrac{MS_{WP}^2}{(w-1)(p-1)} + \dfrac{MS_{WPO}^2}{(w-1)(p-1)(o-1)}}$$

$$= \frac{650^2}{\dfrac{330^2}{27} + \dfrac{640^2}{12} + \dfrac{320^2}{108}} = 10.8$$

$F_{1,11} = 4.015$, $p = .037$. We reject the null hypothesis that the programs have no effects.

Alternatively, we could use Equation 13.19 to calculate

$$F_2' = \frac{MS_P + MS_{WPO}}{MS_{PO} + MS_{WP}} = 3.021. \text{ Then, applying Equations 13.20 and 13.21,}$$

we have $df_{numerator}' = 3.78$ and $df_{denominator}' = 24.65$. $F_{4,25} = 3.021$, $p = .037$, the same result obtained previously.

(c) Note that $F_{27,108} = MS_{WP}/MS_{WPO} = 1.03$, and $p = .44$. Assuming, therefore, that $\sigma^2_{WP} = 0$, $MS_{PO}$ is now an appropriate error term against which to test $P$. The result is $F_{3,12} = 4.078, p < .05$.

**13.9**

(a)

| SV | df | EMS |
|---|---|---|
| Subjects (S) | $n - 1$ | $\sigma^2_e + ot\sigma^2_S$ |
| Occasions (O) | $o - 1$ | $\sigma^2_e + nt\sigma^2_O$ |
| Tasks (T) | $t - 1$ | $\sigma^2_e + no\sigma^2_T$ |
| Residual | $not - (n + o + t - 2)$ | $\sigma^2_e$ |

(b) $\hat{\sigma}^2_S = (MS_S - MS_{res})/ot$; $\hat{\sigma}^2_O = (MS_O - MS_{res})/nt$; $\hat{\sigma}^2_T = (MS_T - MS_{res})/no$.

(c)

| SV | df | EMS |
|---|---|---|
| Subjects (S) | $n - 1$ | $\sigma^2_e + o\sigma^2_{ST} + ot\sigma^2_S$ |
| Occasions (O) | $o - 1$ | $\sigma^2_e + nt\sigma^2_O$ |
| Tasks (T) | $t - 1$ | $\sigma^2_e + o\sigma^2_{ST} + no\sigma^2_T$ |
| S T | $(n - 1)(t - 1)$ | $\sigma^2_e + o\sigma^2_{ST}$ |
| Residual | $(nt - 1)(o - 1)$ | $\sigma^2_e$ |

(b) $\hat{\sigma}^2_S = (MS_S - MS_{ST})/ot$; $\hat{\sigma}^2_O = (MS_O - MS_{res})/nt$; $\hat{\sigma}^2_T = (MS_T - MS_{ST})/no$.

**13.10**

(a) (i) $\mu_1 - (1/3)(\mu_2 + \mu_3 + \mu_4) = 0$. After converting each subject's scores to a single contrast score, the mean contrast is $\hat{\psi} = 2.867$ and $SE(\hat{\psi})$

$= (2.063 / \sqrt{5}) = .923$. Therefore, $t_4 = \hat{\psi} / SE(\hat{\psi}) = 3.107; p = .036$. (ii) CI $=$
$2.867 \pm (2.776)(.923) = .305, 5.429$.

(b), (c) The results of the trend analysis are

**Tests of Within-Subjects Contrasts**

**Measure: MEASURE_1**

| Source | TIME | Type III Sum of Squares | df | Mean Square | F | Sig. |
|--------|------|------------------------|-----|-------------|------|------|
| TIME | Linear | 37.210 | 1 | 37.210 | 25.930 | .007 |
| | Quadratic | .450 | 1 | .450 | .247 | .646 |
| | Cubic | 2.890 | 1 | 2.890 | 2.481 | .190 |
| Error(TIME) | Linear | 5.740 | 4 | 1.435 | | |
| | Quadratic | 7.300 | 4 | 1.825 | | |
| | Cubic | 4.660 | 4 | 1.165 | | |

The $F$ test for curvature (or nonlinearity) is obtained by

$$F_{2,8} = \frac{(MS_{\text{quad}(T)} + MS_{\text{cubic}(T)}) / 2}{(MS_{S \times \text{quad}(T)} + MS_{S \times \text{cubic}(T)}) / 2} = 1.12; p = .41.$$ The best fitting straight line

has a slope significantly different from zero, but the departure from that line is not
significant.

**13.11**

(a)

| SV | df | SS | MS | Error | F |
|----|----|-----|-----|-------|-----|
| $S$ | 4 | 34.467 | 8.167 | | |
| $A$ | 1 | 112.133 | 112.133 | $SA$ | 30.170 |
| $B$ | 2 | 9.800 | 4.900 | $SB$ | 15.474 |
| $AB$ | 2 | 4.067 | 2.033 | $SAB$ | 0.859 |
| $SA$ | 4 | 14.867 | 3.717 | | |
| $SB$ | 8 | 2.533 | 0.317 | | |
| $SAB$ | 8 | 18.933 | 2.367 | | |

(b)

| SV | df | SS | MS | F |
|----|----|----|----|----|
| S | 4 | 11.489 | 2.872 | |
| A | 1 | 112.133 | 37.378 | 30.170 |
| SA | 4 | 14.867 | 1.239 | |

Note that the *MS* and *SS* are one-third of their original values because the "scores" in part (b) are an average of the three scores in part (a) The *F* ratio for *A* is unchanged. Therefore, as long as *B* has fixed effects, averaging over its levels will not change the test of the effects of interest.

(c) The *EMS* are represented in the table. The terms in parentheses do not contribute to the variability in the data when *B* has fixed effects.

| SV | EMS |
|----|-----|
| S | $\sigma_e^2 + (a\sigma_{SB}^2) + ab\sigma_S^2$ |
| A | $\sigma_e^2 + b\sigma_{SA}^2 + (n\sigma_{AB}^2) + (\sigma_{SAB}^2) + nb\theta_A^2$ |
| B | $\sigma_e^2 + a\sigma_{SB}^2 + na\sigma_B^2$ |
| SA | $\sigma_e^2 + \sigma_{SAB}^2 + b\sigma_{SA}^2$ |
| SB | $\sigma_e^2 + a\sigma_{SB}^2$ |
| AB | $\sigma_e^2 + \sigma_{SAB}^2 + n\sigma_{AB}^2$ |
| SAB | $\sigma_e^2 + \sigma_{SAB}^2$ |

(d) When the *B* effects are random, we require a quasi-*F* test of *A*. We can calculate either $F_1'$ or $F_2'$, using Equation 13.17 or Equation 13.19. For example, $F_2' = (MS_A + MS_{SAB})/(MS_{SA} + MS_{SAB}) = 19.91$. From Equations 20.20 and 20.21, the $df = 1.042$ and $5.98$. Therefore, $F_{1,6} = 19.91$; $p = .004$. Note that the *p* value when *B* is viewed as fixed [part (a)] = .0006.

(e) Averaging over the levels of *B* and analyzing the data as if there were only *a* scores for each subject ignores the *AxB* and *SxAxB* variability that contribute to the $A$ mean square if $B$ has random effects. Therefore, in such circumstances, that procedure will generally lead to an inflated Type 1 error rate.

**13.12**
(a) The means are

| $A_1$ | $A_2$ | $A_3$ | $A_4$ |
|---|---|---|---|
| 2304.000 | 2394.750 | 2384.250 | 2418.125 |

The ANOVA results are

| SV | df | SS | MS | F | p | G-G | H-F |
|---|---|---|---|---|---|---|---|
| A | 3 | 59,008.594 | 19669.531 | 2.84 | .062 | 0.097 | 0.076 |
| Error | 21 | 145,010.156 | 6905.246 | | | | |

G-G and H-F are the Greenhouse-Geisser and Huynh-Feldt corrections for nonsphericity; G-G Epsilon = .615; H-F Epsilon = .825.
(b) The means of the Friedman ranks are

| $A_1$ | $A_2$ | $A_3$ | $A_4$ |
|---|---|---|---|
| 1.250 | 2.500 | 2.125 | 3.250 |

The $SS_A = 22.750$ and $\chi_F^2 = SS_A / [a(a+1)/12] = 13.65; p = .003$. An analysis of variance on the ranks gives a similar result:

| SV | df | SS | MS | F | p | G-G | H-F |
|---|---|---|---|---|---|---|---|
| A | 3 | 22.750 | 7.583 | 9.23 | .000 | .003 | .001 |
| Error | 21 | 17.250 | .821 | | | | |

G-G Epsilon = .652; H-F Epsilon .903..
(c) The means of the ranks after all scores have been ranked from 1 to 32 are

| $A_1$ | $A_2$ | $A_3$ | $A_4$ |
|---|---|---|---|
| 12.250 | 17.625 | 17.500 | 18.625 |

The results of the ANOVA on the ranks is

| SV | df | SS | MS | F | p | G-G | H-F |
|---|---|---|---|---|---|---|---|
| A | 3 | 198.750 | 66.250 | 3.02 | .053 | .097 | .081 |
| Error | 21 | 461.250 | 21.964 | | | | |

G-G Epsilon = .534; H-F Epsilon .667.

*Note*: It must be emphasized that when analyzing real data, one test should be selected a priori. Performing all three risks inflating the Type 1 error rate. In the present example, the data do not depart significantly from normality, and the variances are not that different. Therefore, we would choose to do the usual ANOVA. Of course, a better approach would be to collect more data, allowing a better assessment of violations of assumptions, and a more powerful test.

**13.13**

(a) The only negative difference is that for $S_8$. The absolute difference for $S_8$ has a rank of 6. Therefore, the sum of the negative ranks is 6. For $\alpha = .05$, (two-tailed) and $n = 8$, in order to reject $H_0$, $T_- \leq 3$ (see Appendix Table C.10). Therefore, we cannot reject $H_0$.

(b) For each subject, multiply the scores from $A_1$ to $A_4$ by -3, -1, 1, and 3, respectively, and sum the crossproducts to obtain an index of linearity. Only the contrast for $S_7$ is negative. It has the smallest absolute value of the 8 contrasts and therefore has a rank of 1. Because $T_- \leq 3$, we can reject the null hypothesis.

Statistical packages report an approximate $p$ value based on a normal probability approximation. In parts (a) and (b), the reported $p$ values were .093 and .012. The true probabilities are .109 and .016. For example, of $2^8$ patterns of positive and negative ranks, only 4 (all positive, all negative, only the smallest absolute difference positive, only the smallest absolute difference negative) yield $T_+$ or $T_- \leq 1$; therefore, the true (two-tailed) $p$ value in part (b) is $4/236 = .0156$.

**13.14**

(a) An ANOVA of these data yields

| SV | df | SS | MS | F | p | G-G | H-F |
|---|---|---|---|---|---|---|---|
| Period | 2 | 1.633 | .817 | 3.71 | .034 | .036 | .034 |
| S x P | 38 | 8.367 | .220 | | | | |

To calculate Cochran's $Q$ statistic, we first use Equation 13.31 to obtain $MS_{P/S} = (SS_P + SS_{SP})/(df_P + df_{SP})$ 10/40 = .25. Then, by Equation 13.29, $Q = SS_P/MS_{P/S} = 1.633/.25 = 6.532$. $Q$ is distributed approximately as $\chi^2$ on 2 $df$; therefore, $p = .038$, not far from the value obtained by doing a standard ANOVA on the 0/1 data. Note that this test is available in SPSS in the Reliability Analysis module.

**13.15**

(a) The results of the ANOVA are

| SV | SS | df | MS | F | p | G-G | H-F |
|---|---|---|---|---|---|---|---|
| Seasons | 57.989 | 3 | 19.330 | 12.924 | .000 | .000 | .000 |
| Error | 215.372 | 144 | 1.496 | | | | |

G-G Epsilon = .658; H-F Epsilon = .687

$$\text{(b)} \quad \hat{\sigma}^2_{Seasons} = \left(\frac{3}{4}\right)\left(\frac{MS_{Seasons} - MS_{error}}{n}\right) = (.75)(17.834) / 49 = .273.$$

The partial-$\omega^2 = \hat{\sigma}^2_{Seasons} / (\hat{\sigma}^2_{Seasons} + \hat{\sigma}^2_{error}) = .273/(.273 + 1.496) = .154$, a large value.

(c) The results of the trend analysis are:

**Tests of Within-Subjects Contrasts**

**Measure: MEASURE_1**

| Source | SEASONS | Type III Sum of Squares | df | Mean Square | F | Sig. |
|---|---|---|---|---|---|---|
| SEASONS | Linear | 3.629 | 1 | 3.629 | 5.043 | .029 |
| | Quadratic | 48.133 | 1 | 48.133 | 31.725 | .000 |
| | Cubic | 6.227 | 1 | 6.227 | 2.767 | .103 |
| Error(SEASONS) | Linear | 34.539 | 48 | .720 | | |
| | Quadratic | 72.824 | 48 | 1.517 | | |
| | Cubic | 108.009 | 48 | 2.250 | | |

**13.16**

(a) The results of the ANOVA on the transformed scores are

| Source | SS | df | MS | F | P | G-G | H-F |
|---|---|---|---|---|---|---|---|
| Seasons | 9.633 | 3 | 3.211 | 16.694 | .000 | .000 | .000 |
| Error | 27.698 | 144 | .192 | | | | |

G-G Epsilon = .870; H-F Epsilon = .925

$$\text{(b)} \quad \hat{\sigma}^2_{Seasons} = \left(\frac{3}{4}\right)\left(\frac{MS_{Seasons} - MS_{error}}{n}\right) = (.75)(3.019) / 49 = .047.$$

The partial-$\omega^2 = \hat{\sigma}^2_{Seasons} / (\hat{\sigma}^2_{Seasons} + \hat{\sigma}^2_{error}) = .047/(.047 + .192) = .197.$

(c)

**Tests of Within-Subjects Contrasts**

**Measure: MEASURE_1**

| Source | SEASONS | Type III Sum of Squares | df | Mean Square | F | Sig. |
|---|---|---|---|---|---|---|
| SEASONS | Linear | .340 | 1 | .340 | 2.292 | .137 |
| | Quadratic | 8.842 | 1 | 8.842 | 43.524 | .000 |
| | Cubic | .451 | 1 | .451 | 1.999 | .164 |
| Error(SEASONS) | Linear | 7.118 | 48 | .148 | | |
| | Quadratic | 9.751 | 48 | .203 | | |
| | Cubic | 10.829 | 48 | .226 | | |

(d) Before considering other graphs and statistics, note that the epsilon adjustments are much closer to 1.0 for the transformation, indicating a closer approximation to sphericity. This is supported by noting that the polynomial components of the error sum of squares are more nearly equal in the trend analysis of the transformed data. There are other indications that the data on the transformed scale are "cleaner." Boxplots of the original scores exhibit many outliers; there were none in the plots of the transformed data. The plots were also more symmetric, and a comparison of skew and kurtosis statistics for the two data sets also indicates that the transformed distribution is more nearly symmetric, if not normal. Heterogeneity of variance, though still present, was greatly reduced by the transformation; the ratio of largest to smallest variance was about 22 for the original data, but only about 4.6 for the transformed data.. A comparison of the $\omega^2$ values indicates that the effect of seasons is somewhat larger on the transformed scale. Although we must bear in mind that inferences are about the parameters of transformed populations, logarithmic scales are not uncommon, and in many situations may be readily understood. The present example suggests that the assumptions underlying our inferences may be better met by transforming some data sets.

## Chapter 14

**14.1**

(a) The results of the ANOVA are

| SV | df | SS | MS | F | p |
|---|---|---|---|---|---|
| Between Ss | 5 | 226.5 | | | |
| A | 1 | 112.5 | 112.50 | 3.947 | .118 |
| S/A | 4 | 114.0 | 28.50 | | |
| Within Ss | 12 | 758.0 | | | |
| B | 2 | 84.0 | 42.00 | .994 | .412 |
| AB | 2 | 336.0 | 168.00 | 3.976 | .063 |
| SxB/A | 8 | 338.0 | 42.25 | | |

(b) The ANOVA based on the mean scores for the subjects is

| SV | df | SS | MS | F | p |
|---|---|---|---|---|---|
| Total | 5 | 75.5 | | | |
| A | 1 | 37.5 | 37.5 | 3.947 | .118 |
| S/A | 4 | 38.0 | 9.5 | | |

(i) The $F$s in parts (a) and (b) are identical. (ii) The $SS$ and $MS$ in part (b) are 1/3 of their counterparts in part (a). The reason for this is that in part (a), $SS_A = bn\sum_j(\bar{Y}._j.-\bar{Y}...)^2$ whereas in part (b), $SS_A = n\sum_j(\bar{Y}._j.-\bar{Y}...)^2$.

(c) $SS_{SB/A_1} = 118$ and $SS_{SB/A_2} = 220$; the sum is 338, the result in part (a). Also, $MS_{SB/A_1} = 29.5$ and $MS_{SB/A_2} = 55$; the average is 42.25, the result in part (a).

**14.2**

(a)

| SV | df | EMS |
|---|---|---|
| A | 1 | $\sigma_e^2 + 3\sigma_{S/A}^2 + 9\sigma_A^2$ |
| S/A | 4 | $\sigma_e^2 + 3\sigma_{S/A}^2$ |
| B | 2 | $\sigma_e^2 + \sigma_{SB/A}^2 + 3\sigma_{AB}^2 + 6\theta_B^2$ |
| AB | 2 | $\sigma_e^2 + \sigma_{SB/A}^2 + 3\sigma_{AB}^2$ |
| SB/A | 8 | $\sigma_e^2 + \sigma_{SB/A}^2$ |

(b) $B$ is now tested against $AB$; therefore, $F = 42/168 = .25$, which is clearly not significant.

**14.3**

| SV | df | EMS |
|----|----|-----|
| $A$ | 1 | $\sigma_e^2 + \sigma_{SB/A}^2 + 3\sigma_{S/A}^2 + 3\sigma_{AB}^2 + 9\theta_A^2$ |
| $S/A$ | 4 | $\sigma_e^2 + \sigma_{SB/A}^2 + 3\sigma_{S/A}^2$ |
| $B$ | 2 | $\sigma_e^2 + \sigma_{SB/A}^2 + 6\theta_B^2$ |
| $AB$ | 2 | $\sigma_e^2 + \sigma_{SB/A}^2 + 3\sigma_{AB}^2$ |
| $SB/A$ | 8 | $\sigma_e^2 + \sigma_{SB/A}^2$ |

To test the $A$ effect, we form a quasi-$F$ ratio. One possibility is
$F' = (MS_A + MS_{SB/A})/(MS_{S/A} + MS_{AB}) = (112.5 + 42.25)/(28.5 + 168) = .79$.
Alternatively, we can calculate $F' = MS_A /(MS_{S/A} + MS_{AB} - MS_{SB/A}) = .73$. In either
case, the result is clearly not significant.

**14.4**

(a)

| SV | df |
|----|----|
| $A$ | 2 |
| $X$ | 1 |
| $AX$ | 2 |
| $S/AX$ | 114 |
| $V$ | 2 |
| $AV$ | 4 |
| $XV$ | 2 |
| $AXV$ | 4 |
| $SV/AX$ | 228 |

(b)

| SV | df |
|----|----|
| $A$ | 2 |
| $X$ | 1 |
| $C$ | 1 |
| $AX$ | 2 |
| $AC$ | 2 |
| $XC$ | 1 |
| $AXC$ | 2 |
| $S/AXC$ | 108 |
| $V$ | 2 |

| | | | |
|-----|-----|--------|-----|
| *C* | 1 | *AV* | 4 |
| *AC* | 2 | *XV* | 2 |
| *XC* | 1 | *VC* | 2 |
| *AXC* | 2 | *AVC* | 4 |
| *SC/AX* | 114 | *AXV* | 4 |
| *VC* | 2 | *XVC* | 2 |
| *AVC* | 4 | *AXVC* | 4 |
| *XVC* | 2 | *SV/AXC* | 216 |
| *AXVC* | 4 | | |
| *SVC/AX* | 228 | | |

In the design of part (b), the *C, AC, XC,* and *AXC* terms are tested against a between-subjects error term. Because of the increased variability due to individual differences, these tests are likely to have less power than in the design of part (a). Furthermore, interval estimates of the difference in the *C* means, and of interactions with *A* and *X* will be less precise; the intervals will be wider.

(c) The advantage of the design is that it requires less time for each subject, which may be an important consideration if each observation requires time to obtain (for example, if each response was based on listening to a story, or viewing a segment of a TV program), or if exposure to one level of a variable will affect responses at another level (which may well be true with levels of violence). The disadvantage is the need for very large numbers of subjects to compensate for the likely large error variance due to individual differences.

**14.5** (a) The ANOVA table is

| SV | df | EMS |
|---|---|---|
| Between Ss | 71 | |
| X | 1 | $\sigma_e^2 + 2\sigma_{S/XA}^2 + 72\theta_X^2$ |
| A | 2 | $\sigma_e^2 + 2\sigma_{S/XA}^2 + 48\theta_A^2$ |
| XA | 2 | $\sigma_e^2 + 2\sigma_{S/XA}^2 + 24\theta_{XA}^2$ |
| S/XA | 66 | $\sigma_e^2 + 2\sigma_{S/XA}^2$ |
| Within Ss | 72 | |
| T | 1 | $\sigma_e^2 + \sigma_{ST/XA}^2 + 72\theta_T^2$ |
| XT | 1 | $\sigma_e^2 + \sigma_{ST/XA}^2 + 36\theta_{XT}^2$ |
| AT | 2 | $\sigma_e^2 + \sigma_{ST/XA}^2 + 24\theta_{AT}^2$ |
| XAT | 2 | $\sigma_e^2 + \sigma_{ST/XA}^2 + 12\theta_{XAT}^2$ |
| ST/XA | 66 | $\sigma_e^2 + \sigma_{ST/XA}^2$ |

*S/XA* is the error term for the between-subjects terms, and *ST/XA* is the error term for the within-subjects terms. However, note that in a study like this, the investigator would most likely counterbalance the order of the two tasks. This would result in order (*O*), an additional between-subjects, fixed-effect variable with two levels, being added to the design.

(b) The most conservative approach is to base the error term only on those scores involved in these tests of simple effects; that is the default in most statistical packages. For part (i), we find the variance of $T_1$ scores within each age x sex combination, and average these. We might indicate this as $MS_{S/AxX/T_1}$ (subjects within age x sex combinations for the $T_1$ task). The $df = ax(n-1) = (3)(2)(11) = 66$.

(ii) The error term is $MS_{S/A/\text{Male}/T_1}$. The $df = a(n-1) = (3)(11) = 33$.

(iii) The error term is $MS_{ST/A/\text{Male}}$. The $df = a(n-1)(t-1) = (3)(11)(1) = 33$.

**14.6**

(a) ) $\hat{f} = \sqrt{(df_{\text{groups}} / N)(F_{\text{groups}} - 1)} = \sqrt{(1 / 48)(5.20)} = .333;$

$\hat{\omega}^2 = \hat{f}^2 / (\hat{f}^2 + 1) = .098.$   Cohen suggests that $f = .25$ is medium and .4 is large.  This effect size is about halfway between these values.

(b) Using GPOWER, we enter a value for $f^2$ of .108.  Using SPSS or SAS, or the UCLA calculator, we enter the noncentrality parameter, $\lambda = Nf^2 = 5.184.$ In either case, power = .58.  Nonsphericity is not a problem here because we are dealing with a between-subjects effect.  It also would not be a problem for assessing the power to test Conditions or its interactions because these – although within-subjects terms – are distributed on 1 $df$.  In essence, we are dealing with a single set of difference scores, so that heterogeneity of differences is not an issue.

(c) For a medium effect size $f^2 = .0625$ and $\lambda = .0625N$.  We found that $N = 140$ yielded power = .8033.  Because there are two scores for each subject, the number of subjects would be 70.  However, because there are four groups $x$ order combinations, we would require 72 subjects, or $n = 18$.

**14.7** The confidence interval is $CI = \hat{\psi} \pm SE(\hat{\psi}) \cdot t_{.05,df} \cdot \hat{\psi} = $ $(14.8 - 9.8) - (11 - 11.4) = 5.4.$  Because testing the contrast is equivalent to testing the interaction, we can calculate $SE(\hat{\psi})$ from $t = \sqrt{F} = \sqrt{16.89} = $ 4.11.  But $t = $ $\hat{\psi} / SE(\hat{\psi});$ solving, $SE(\hat{\psi}) = 5.4/4.11 = 1.314.$  The within-subjects error term is on $4(6-1)(2-1)$ $df$, or 20 $df$ and $t_{.05,20} = 2.086.$  Therefore the confidence interval is CI = $5.4 \pm (1.314)(2.086) = 2.66, 8.14.$

**14.8**   (a)   The plot of Diet $x$ Days:

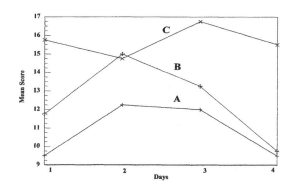

(b) The results of the ANOVA are

| SV | df | SS | MS | F | p |
|---|---|---|---|---|---|
| Diet (D) | 2 | 197.167 | 98.583 | 13.637 | .002 |
| S/D | 9 | 65.063 | 7.229 | | |
| Days (d) | 3 | 53.396 | 17.799 | 12.923 | .000 |
| lin(d) | 1 | 3.037 | 3.037 | 1.153 | .311 |
| quad(d) | 1 | 50.021 | 50.021 | 74.258 | .000 |
| cubic(d) | 1 | .337 | .337 | .410 | .538 |
| D x d | 6 | 42.167 | 7.028 | 5.103 | .001 |
| lin(d) x D | 2 | 9.300 | 4.650 | 1.765 | .226 |
| quad(d) x D | 2 | 23.167 | 11.583 | 17.196 | .001 |
| cubic(d) x D | 2 | 9.700 | 4.850 | 5.889 | .023 |
| S x d /D | 27 | 37.188 | 1.377 | | |
| S x lin(d)/D | 9 | 23.713 | 2.635 | | |
| S x quad(d)/D | 9 | 6.062 | .674 | | |
| S x cubic(d)/D | 9 | 7.413 | .824 | | |

The main effect of diet allows us to conclude that, averaging over days, the diet population means differ. The quad(d) term, coupled with the clear lack of significance for linear and cubic components of Days, suggests that the average population function is adequately described by an inverted $U$-shaped curve. The components of the interaction $(D \times d)$ indicate that the main differences among the shapes of the three curves lies in their curvature. The best-fitting straight lines are nearly flat, accounting for the lack of a lin(d) x D effect. The significant cubic and quadratic interaction components reflect the fact that the curve for $C$ is essentially flat whereas the other two diets yield an initial improvement followed by a performance drop on the last day.

**14.9**

(a) Entering Appendix Table C.10 with 9 df (the df for the S/D term in the ANOVA) and three ordered means, the critical studentized range statistic is $q_{.05,9}$

= 3.95. The standard error of the mean can be obtained by dividing $MS_{S/D}$ by $bn$ (= 16), and taking the square root; this is equivalent to finding the mean for each subject, calculating the variances of these means about their respective group mans, averaging the variances, dividing by $n$, and taking the square root. The result is .672. The confidence interval is $\hat{\psi} \pm q_{.05,9} SE(\overline{Y}_{i}..)$ or $\hat{\psi} \pm (3.95)(.672)$. Accordingly, the B-A interval is $(12.438 - 10.813) \pm 2.654 = -1.03, 4.28$; the C - A interval is $(15.688 - 10.813) \pm 2.654 = 2.22, 7.53$; and the C-B interval is $(15.688 - 12.438) \pm 2.654 = .60, 5.90$. Diet C differs significantly from both Diets A and B; Diets A and B do not differ significantly from each other.

(b) Scheffé's method is appropriate here. The confidence interval formula is $CI = \hat{\psi} \pm S_{\hat{\psi}} \sqrt{df_D \cdot F_{.05,2,9}}$ where $\hat{\psi} = 15.5 - (1/2)(9.5 + 9.75) = 5.875$,

$S_{\hat{\psi}} = \sqrt{\left(\sum_j w_j^2 / n\right) MS_{S/D/Day4}}$, the $df$ for Diet = 2, and the critical value of $F$ is 4.26. To get the SE of the contrast, we perform an ANOVA using only the Day 4 data. The error mean square is 4.528 and $s_{\hat{\psi}} = \sqrt{(1.5)/4)(4.528)} = 1.303$.

Therefore, the confidence interval is $CI = 5.875 \pm 1.303\sqrt{(2)(4.26)} = 2.07, 9.68$.

Since the interval does not contain zero, we conclude that the mean for diet C does differ from the average of the other two means on Day 4.

**14.10**

(a) $F_{1,6} = 80.083/8.750 = 9.152, p = .023$.

(b) We must now form quasi-$F$ ratios to test $A$, $B$, and $AB$. To test the $A$ source, $F' = (MS_A + MS_{SC/A})/(MS_{S/A} + MS_{AC}) = (80.083 + 2.271)/(8.750 + 21.271) = 2.74$. The numerator and denominator $df$ are

$$df_1 = \frac{(MS_A + MS_{SC/A})^2}{(MS_A^2 / df_A) + (MS_{SC/A}^2 / df_{SC/A})} = \frac{(80.083 + 2.271)}{(80.083^2 / 1) + (2.271^2 / 12)} = 1.06$$

and $$df_2 = \frac{(MS_{S/A} + MS_{AC})^2}{(MS_{S/A}^2 / df_{S/A}) + (MS_{AC}^2 / df_{AC})} = \frac{(8.750 + 21.271)}{(8.750^2 / 6) + (21.271^2 / 2)} = 3.78.$$

The $df$ are rounded to 1 and 4 and $p = .17$. Note that when we take into consideration the variance due to the random factor $C$, the $p$ value is considerably increased. If we wish to generalize to the sampled population of items, the result is no longer significant.

(c) For $B$, $F' = (MS_B + MS_{SBC/A})/(MS_{SB/A} + MS_{BC})$; for $AB$, $F' = (MS_{AB} + MS_{SBC/A})/(MS_{SB/A} + MS_{AB})$.

**4.11**

(a) The figure :

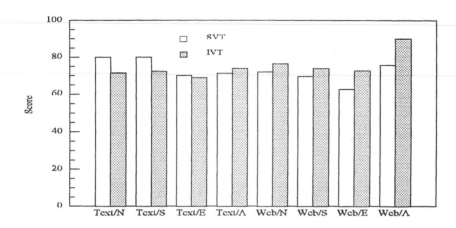

Mean SVT scores are higher than IVT scores in the Text format, though the advantage diminishes as we move from N (narrative) to A (argument) instructions. In the web format, IVT scores are higher than SVT scores and the advantage is greatest with A instructions. This suggests a Test $x$ Format interaction, and a Test $x$ Instruction interaction.

(b)

| SV | df | SS | MS | F | p |
|---|---|---|---|---|---|
| Format ($F$) | 1 | 9.570 | 9.570 | .05 | .818 |
| Instructions($I$) | 3 | 1430.273 | 476.758 | 2.67 | .056 |
| FxI | 3 | 1064.648 | 354.883 | 1.99 | .126 |
| S/FI | 56 | 9995.312 | 178.488 | | |
| Test ($T$) | 1 | 164.258 | 164.258 | 1.67 | .202 |
| TxF | 1 | 1158.008 | 1158.008 | 11.75 | .001 |
| TxI | 3 | 616.211 | 205.404 | 2.08 | .113 |
| TxFxI | 3 | 3.711 | 1.237 | .01 | .998 |

The one source that is significant at the .05 level is the Test $x$ Format interaction (TxF); $F_{1,56}$ =11.75, p = .001. This interaction reflects the fact that the sentence memory (SVT) mean is higher when participants have studied the material from a textbook chapter but that the inferences(IVT) mean is higher when participants have been required to integrate information from the website.

**14.12**

(a) The simplest way to proceed is to construct a new dependent variable, d = IVT-SVT, and conduct a two-factor (format, instructions) ANOVA. Then $\hat{f} = \sqrt{(df_{Format} / N)(F-1)} = \sqrt{(1/64)(10.747)} = .41$, a large effect under Cohen's guidelines.

(b) Subtracting the SVT mean from the IVT mean, we have -3.750 for the text format and 8.281 for the web format. Therefore, the interaction effect is 8.281 - (-3.750) = 12.031. To obtain a confidence interval, think of it as an interval containing the difference between the means of the text and web populations of difference scores. As usual, the general form of the confidence interval is $\overline{D} \pm t s_{\overline{D}}$, where $\overline{D}$ is the difference between the text and web mean differences, or 12.031, $s_{\overline{D}}$ is the standard error of this interaction effect, and $t$ is the value that cuts of .025 in each tail of the $t$ distribution. The standard error is $s_{\overline{D}} = \sqrt{(2/4n)MS_{error}}$. The "2" is because we are working with the variance of the difference between two means, and $4n$ is the number of scores on which each mean is based. The error mean square is obtained from the ANOVA of the IVT - SVT difference scores. Therefore, $s_{\overline{D}} = \sqrt{(2/32)(197.154)} = 3.510$. The critical value of $t$ is based on the error $df$, $t_{.05,56} \approx 2.00$. Therefore, $CI = 12.031 \pm (3.501)(2.00) = 5.03, 19.03$.

(c) Perform two ANOVAs on the difference scores calculated in part (a), one in each format condition. (i) In the text condition, $\hat{f} = \sqrt{(df_I / N_{text})(F-1)}$ $= \sqrt{(3/32)(.110)} = .102$, a small effect. (ii) Because $F < 1$ in the web condition, our best estimate of $f$ is zero. In either case, instructions have very little effect, relative to the error variance, on the difference between the two measures.

## Chapter 15

**15.1**

(a) The effect of $P$ is not significant when an ANOVA is performed.

```
Dep Var: Y   N: 36   Multiple R: 0.344   Squared multiple R: 0.119
Analysis of Variance
```

| Source | Sum-of-Squares | df | Mean-Square | F-ratio | P |
|--------|----------------|-----|-------------|---------|------|
| P | 806.167 | 2 | 403.083 | 2.221 | 0.124 |
| Error | 5988.583 | 33 | 181.472 | | |

(b) Using an ANCOVA, the effect of $P$ is significant.

```
Dep Var: Y   N: 36   Multiple R: 0.830   Squared multiple R: 0.688
Analysis of Variance
```

| Source | | Sum-of-Squares | df | Mean-Square | F-ratio | P |
|--------|--------|----------------|-----|-------------|---------|------|
| X | | 3869.741 | 1 | 3869.741 | 58.443 | 0.000 |
| P | [P(adj)] | 539.495 | 2 | 269.748 | 4.074 | 0.027 |
| Error | [S/P (adj)] | 2118.843 | 32 | 66.214 | | |

(c) We use the GLM module in one of the software packages to add a $P*X$ term to the model. The interaction term is not significant, so we don't reject the hypothesis of homogeneity of slopes.

```
Dep Var: Y   N: 36   Multiple R: 0.844   Squared multiple R: 0.712
Analysis of Variance
Source  Sum-of-Squares  df   Mean-Square  F-ratio  P
X       3725.614         1   3725.614     57.130   0.000
P         75.463         2     37.731      0.579   0.567
P*X      162.458         2     81.229      1.246   0.302
Error   1956.385        30     65.213
```

### 15.2

(a) The ANOVA on $Y$ does not yield a significant result.

```
Dep Var: Y   N: 18   Multiple R: 0.453   Squared multiple R: 0.205
Analysis of Variance
Source   Sum-of-Squares  df   Mean-Square  F-ratio   P
A         172.111         2    86.056       1.936    0.179
Error     666.833        15    44.456
```

(b) There is no evidence that the population means for $X$ differ. In fact, the sample means are less variable than we would expect by chance.

```
Dep Var: X   N: 18   Multiple R: 0.065   Squared multiple R: 0.004
Analysis of Variance
Source  Sum-of-Squares  df   Mean-Square  F-ratio  P
A          0.778         2    0.389        0.032    0.969
Error    184.167        15   12.278
```

(c) There is no evidence that the assumption of homogeneity of regression slopes is violated.

```
Dep Var: Y   N: 18   Multiple R: 0.974   Squared multiple R: 0.948
Analysis of Variance
Source  Sum-of-Squares  df   Mean-Square  F-ratio   P
A         10.128         2     5.064        1.402   0.284
X        503.690         1   503.690      139.468   0.000
A*X        2.554         2     1.277        0.354   0.709
Error     43.338        12     3.612
```

(d) The ANCOVA on $Y$ using $X$ as a covariate indicates a significant $A$ effect.

```
Dep Var: Y   N: 18   Multiple R: 0.972   Squared multiple R: 0.945
Analysis of Variance
Source  Sum-of-Squares  df   Mean-Square  F-ratio   P
A        182.822         2    91.411       27.886   0.000
X        620.941         1   620.941      189.427   0.000
Error     45.892        14     3.278
```

(e) Because here subjects are assigned randomly to groups, the hypotheses tested test by performing an ANOVA on the $Y$ scores and by performing an ANCOVA on the $Y$ scores using $X$ as the covariate are the same: the population means for $Y$ are identical for the different levels of $A$.

(f) The adjusted means are 26.102, 30.157, and 33.908.

(g) The adjusted mean for a group is the $Y$ score predicted using the within-group regression equation of $Y$ on $X$ with the common slope, if $X$ is equal to the grand mean of the $X$ scores. They are estimates of what the group means would be if all $X$ scores were equal to the grand mean.

**15.3**

(a) No, it is not appropriate to use ANCOVA here. We have a nonequivalent-groups design because the workers for whom we have satisfaction scores have not been randomly assigned to the four departments. Moreover, an ANOVA with $X$ as the dependent variable yields a significant effect of department, $F(3, 28) = 5.602, p = .004$.

(b) No, it is not appropriate to use ANCOVA here. The data violate the assumption of homogeneity of regression slopes. The test of heterogeneity of slopes indicates that there is a significant interaction between the covariate $X$ and the factor $A$, $F(2, 24) = 7.137, p = .004$.

**15.4**
```
Dep Var: Y    N: 18    Multiple R: 0.894    Squared multiple R: 0.799
Analysis of Variance
```

| Source | Sum-of-Squares | df | Mean-Square | F-ratio | P |
|---|---|---|---|---|---|
| A | 182.755 | 2 | 91.377 | 29.821 | 0.000 |
| Error | 45.963 | 15 | 3.064 | | |

The ANOVA on the residuals is not equivalent to an ANCOVA, although the differences are small for this data set. ANCOVA involves testing the full model
$$Y_{ij} = \mu + \alpha_j + \beta(X_{ij} - \overline{X}..) + \varepsilon$$
against the restricted model
$$Y_{ij} = \mu + \beta(X_{ij} - \overline{X}..) + \varepsilon$$
where $\beta$ in the restricted model is the slope of the overall regression of $Y$ on $X$ estimated by $b_{tot}$ and $\beta$ in the full model is estimated by $b_{S/A}$. Using the residuals from the overall regression is equivalent to testing
$$Y_{ij} = \mu + \alpha_j + \beta(X_{ij} - \overline{X}..) + \varepsilon$$
against
$$Y_{ij} = \mu + \beta(X_{ij} - \overline{X}..) + \varepsilon$$
where, in both models, $\beta$ is the slope of the overall regression of $Y$ on $X$.

**15.5** The ANOVA on $Y$ does not indicate any significant effects; however, the ANCOVA on $Y$ using $X$ as a covariate reveals a significant $A$ effect, $F(1, 19) = 17.376, p = .001$. There are no significant effects when an ANOVA is performed using $X$ as the dependent variable; also, there is no suggestion that the homogeneity of regression slopes has been violated.

```
Dep Var: Y    N: 24    Multiple R: 0.409 Squared multiple R:   0.168
Analysis of Variance
```

| Source | Sum-of-Squares | df | Mean-Square | F-ratio | P |
|---|---|---|---|---|---|
| A | 118.815 | 1 | 118.815 | 3.963 | 0.060 |
| B | 1.215 | 1 | 1.215 | 0.041 | 0.842 |
| A*B | 0.667 | 1 | 0.667 | 0.022 | 0.883 |
| Error | 599.577 | 20 | 29.979 | | |

```
Dep Var: Y    N: 24    Multiple R: 0.762    Squared multiple R: 0.581
Analysis of Variance
```

| Source | Sum-of-Squares | df | Mean-Square | F-ratio | P |
|---|---|---|---|---|---|
| A | 276.023 | 1 | 276.023 | 17.376 | 0.001 |
| B | 2.513 | 1 | 2.513 | 0.158 | 0.695 |
| A*B | 11.878 | 1 | 11.878 | 0.748 | 0.398 |
| X | 297.762 | 1 | 297.762 | 18.745 | 0.000 |
| Error | 301.814 | 19 | 15.885 | | |

```
Dep Var: X    N: 24    Multiple R: 0.421    Squared multiple R: 0.177
Analysis of Variance
```

| Source | Sum-of-Squares | df | Mean-Square | F-ratio | P |
|---|---|---|---|---|---|
| A | 64.354 | 1 | 64.354 | 3.334 | 0.083 |
| B | 9.500 | 1 | 9.500 | 0.492 | 0.491 |
| A*B | 9.250 | 1 | 9.250 | 0.479 | 0.497 |
| Error | 386.082 | 20 | 19.304 | | |

# Chapter 16

**16.1**

(a)

| SV | df | EMS |
|---|---|---|
| $D$ | 2 | $\sigma_e^2 + 20\sigma_{C/D}^2 + 100\theta_D^2$ |
| $C/D$ | 12 | $\sigma_e^2 + 20\sigma_{C/D}^2$ |
| $X$ | 1 | $\sigma_e^2 + 10\sigma_{CX/D}^2 + 150\theta_X^2$ |
| $DX$ | 2 | $\sigma_e^2 + 10\sigma_{CX/D}^2 + 50\theta_{DX}^2$ |
| $CX/D$ | 12 | $\sigma_e^2 + 10\sigma_{CX/D}^2$ |
| $S/CX/D$ | 270 | $\sigma_e^2$ |

(b) $MS_{S/DX}$ is the pool of the $C/D$, $CX/D$, and $S/CX/D$ mean squares; $E(MS_{S/DX}) = \sigma_e^2 + (12/294)(20\sigma_{C/D}^2) + (12/294)(10\sigma_{CX/D}^2)$. If the variance due to $C/D > 0$, the test of $X$ will be negatively biased; if the variance due to $CX/D > 0$, the test of $D$ will be negatively biased.

**16.2**

| SV | df | EMS | Error Term |
|----|----|-----|-----------|
| $T$ | 1 | $\sigma_e^2 + 6\sigma_{G/T}^2 + 36\theta_T^2$ | $G/T$ |
| $G/T$ | 10 | $\sigma_e^2 + 6\sigma_{G/T}^2$ | $S/GX/T$ |
| $X$ | 1 | $\sigma_e^2 + 3\sigma_{GX/T}^2 + 36\theta_X^2$ | $GX/T$ |
| $TX$ | 1 | $\sigma_e^2 + 3\sigma_{GX/T}^2 + 18\theta_{TX}^2$ | $GX/T$ |
| $GX/T$ | 10 | $\sigma_e^2 + 3\sigma_{GX/T}^2$ | $S/GX/T$ |
| $S/GX/T$ | 48 | $\sigma_E^2$ | |

**16.3** Presumably, we wish to generalize beyond the four leaders used in the study. Therefore $L$ is viewed as a random-effects variable. The ANOVA is

| SV | df | EMS | Error Term |
|----|----|-----|-----------|
| $L$ | 3 | $\sigma_e^2 + 6\sigma_{G/LM}^2 + 60\theta_L^2$ | $G/LM$ |
| $M$ | 1 | $\sigma_e^2 + 6\sigma_{G/LM}^2 + 30\sigma_{LM}^2 + 120\theta_M^2$ | $LM$ |
| $LxM$ | 3 | $\sigma_e^2 + 6\sigma_{G/LM}^2 + 30\sigma_{LM}^2$ | $G/LM$ |
| $G/LM$ | 32 | $\sigma_e^2 + 6\sigma_{G/LM}^2$ | $S/G/LM$ |
| $S/G/LM$ | 200 | $\sigma_e^2$ | |

If each leader has been selected for a particular quality, we may view $L$ as fixed, in which case $M$ is tested against $G/LM$.

**16.4**

(a)

| SV | df | EMS | Error Term |
|---|---|---|---|
| $P$ | 1 | $\sigma_e^2 + 2\sigma_{S/C/Sc/P}^2 + 20\sigma_{C/Sc/P}^2 + 40\sigma_{Sc/P}^2 + 120\theta_P^2$ | $Sc/P$ |
| $Sc/P$ | 4 | $\sigma_e^2 + 2\sigma_{S/C/Sc/P}^2 + 20\sigma_{C/Sc/P}^2 + 40\sigma_{Sc/P}^2$ | $C/Sc/P$ |
| $C/Sc/P$ | 6 | $\sigma_e^2 + 2\sigma_{S/C/Sc/P}^2 + 20\sigma_{C/Sc/P}^2$ | $S/C/Sc/P$ |
| $S/C/Sc/P$ | 108 | $\sigma_e^2 + 2\sigma_{S/C/Sc/P}^2$ | |
| $T$ | 1 | $\sigma_e^2 + \sigma_{ST/C/Sc/P}^2 + 10\sigma_{CT/Sc/P}^2 + 20\sigma_{ScT/P}^2 + 120\theta_T^2$ | $ScT/P$ |
| $PT$ | 1 | $\sigma_e^2 + \sigma_{ST/C/Sc/P}^2 + 10\sigma_{CT/Sc/P}^2 + 20\sigma_{ScT/P}^2 + 60\theta_{PT}^2$ | $ScT/P$ |
| $ScT/P$ | 4 | $\sigma_e^2 + \sigma_{ST/C/Sc/P}^2 + 10\sigma_{CT/Sc/P}^2 + 20\sigma_{ScT/P}^2$ | $CT/Sc/P$ |
| $CT/Sc/P$ | 6 | $\sigma_e^2 + \sigma_{ST/C/Sc/P}^2 + 10\sigma_{CT/Sc/P}^2$ | $ST/C/Sc/P$ |
| $ST/C/Sc/P$ | 108 | $\sigma_e^2 + \sigma_{ST/C/Sc/P}^2$ | |

(b) This design provides no variability due to classes. Assuming that the variance due to classes is zero, the ANOVA is

| SV | df | EMS | Error Term |
|---|---|---|---|
| $P$ | 1 | $\sigma_e^2 + 2\sigma_{S/ScP}^2 + 20\sigma_{ScP}^2 + 120\theta_P^2$ | $ScP$ |
| $Sc$ | 5 | $\sigma_e^2 + 2\sigma_{S/ScP}^2 + 40\sigma_{Sc}^2$ | $S/ScP$ |
| $ScP$ | 5 | $\sigma_e^2 + 2\sigma_{S/ScP}^2 + 20\sigma_{ScP}^2$ | $S/ScP$ |
| $S/ScP$ | 108 | $\sigma_e^2 + 2\sigma_{S/ScP}^2$ | |
| $T$ | 1 | $\sigma_e^2 + \sigma_{ST/ScP}^2 + 20\sigma_{ScT}^2 + 120\theta_T^2$ | $ScT$ |
| $PT$ | 1 | $\sigma_e^2 + \sigma_{ST/ScP}^2 + 10\sigma_{ScPT}^2 + 60\theta_T^2$ | $ScPT$ |
| $ScT$ | 5 | $\sigma_e^2 + \sigma_{ST/ScP}^2 + 20\sigma_{ScT}^2$ | $ST/ScP$ |
| $ScPT$ | 5 | $\sigma_e^2 + \sigma_{ST/ScP}^2 + 10\sigma_{ScPT}^2$ | $ST/ScP$ |
| $ST/ScP$ | 108 | $\sigma_e^2 + \sigma_{ST/ScP}^2$ | |

**16.5**

| SV | df | EMS |
|---|---|---|
| S | 19 | $\sigma_e^2 + 50\sigma_S^2$ |
| M | 4 | $\sigma_e^2 + 20\sigma_{I/M}^2 + 10\sigma_{SM}^2 + 200\theta_M^2$ |
| I / M | 45 | $\sigma_e^2 + 20\sigma_{I/M}^2$ |
| SxM | 76 | $\sigma_e^2 + 10\sigma_{SM}^2$ |
| SxI /M | 855 | $\sigma_e^2$ |

To test $M$, a quasi-$F$ is needed; $F' = (MS_M + MS_{SI/M})/(MS_{I/M} + MS_{SM})$. The numerator $df = (MS_M + MS_{SI/M})^2 / (MS_M^2 / 4 + MS_{SI/M}^2 / 855)$; the denominator $df = (MS_{I/M} + MS_{SM})^2 / (MS_{I/M}^2 / 45 + MS_{SM}^2 / 76)$.

**16.6**

| SV | df | EMS |
|---|---|---|
| A | 2 | $\sigma_e^2 + 15\sigma_{S/A}^2 + 10\sigma_{AE/V}^2 + 150\theta_A^2$ |
| S/A | 27 | $\sigma_e^2 + 15\sigma_{S/A}^2$ |
| V | 2 | $\sigma_e^2 + 5\sigma_{VS/A}^2 + 30\sigma_{E/V}^2 + 150\theta_V^2$ |
| AV | 4 | $\sigma_e^2 + 5\sigma_{VS/A}^2 + 10\sigma_{AE/V}^2 + 50\theta_{AV}^2$ |
| VS/A | 54 | $\sigma_e^2 + 5\sigma_{VS/A}^2$ |
| E/V | 12 | $\sigma_e^2 + 30\sigma_{E/V}^2$ |
| AE/V | 24 | $\sigma_e^2 + 10\sigma_{AE/V}^2$ |
| SE/AV | 324 | $\sigma_e^2$ |

To test the $A$ source of variance, $F' = (MS_A + MS_{SE/AV})/(MS_{S/A} + MS_{AE/V})$. The numerator $df = (MS_A + MS_{SE/AV})^2 / (MS_A^2 / 2 + MS_{SE/AV}^2 / 324)$ and the denominator $df = (MS_{S/A} + MS_{AE/V})^2 / (MS_{S/A}^2 / 27 + MS_{AE/V}^2 / 24)$.

**16.7**

| SV | df | EMS |
|----|----|-----|
| A | 2 | $\sigma_e^2 + 5\sigma_{S/AV}^2 + 10\sigma_{AE/V}^2 + 150\theta_A^2$ |
| V | 2 | $\sigma_e^2 + 5\sigma_{S/AV}^2 + 30\sigma_{E/V}^2 + 150\theta_V^2$ |
| AV | 4 | $\sigma_e^2 + 5\sigma_{S/AV}^2 + 10\sigma_{AE/V}^2 + 50\theta_{AV}^2$ |
| S/AV | 81 | $\sigma_e^2 + 5\sigma_{S/AV}^2$ |
| E/V | 12 | $\sigma_e^2 + 30\sigma_{E/V}^2$ |
| AE/V | 24 | $\sigma_e^2 + 10\sigma_{AE/P}^2$ |
| SE/AV | 324 | $\sigma_e^2$ |

To test the $A$ source of variance, $F' = (MS_A + MS_{SE/AV})/(MS_{S/AV} + MS_{AE/V})$. The numerator $df = (MS_A + MS_{SE/AV})^2 / (MS_A^2 / 2 + MS_{SE/AV}^2 / 324)$ and the denominator $df = (MS_{S/AV} + MS_{AE/V})^2 / (MS_{S/AV}^2 / 81 + MS_{AE/V}^2 / 24)$.

**16.8** Let $A$ = age, $E$ = environment, $P$ = problem, and $G$ = group.

| SV | df | EMS |
|----|----|-----|
| A | 1 | $\sigma_e^2 + 8\sigma_{S/G/A}^2 + 24\sigma_{G/A}^2 + 3\sigma_{GP/AE}^2 + 15\sigma_{AP/E}^2 + 120\theta_A^2$ |
| G/A | 8 | $\sigma_e^2 + 8\sigma_{S/G/A}^2 + 24\sigma_{G/A}^2 + 3\sigma_{GP/AE}^2$ |
| S/G/A | 20 | $\sigma_e^2 + 8\sigma_{S/G/A}^2$ |
| E | 1 | $\sigma_e^2 + 12\sigma_{EG/A}^2 + 4\sigma_{SE/G/A}^2 + 3\sigma_{GP/AE}^2 + 30\sigma_{P/E}^2 + 120\theta_E^2$ |
| AE | 1 | $\sigma_e^2 + 12\sigma_{EG/A}^2 + 4\sigma_{SE/G/A}^2 + 3\sigma_{GP/AE}^2 + 15\sigma_{AP/E}^2 + 60\theta_{AE}^2$ |
| EG/A | 8 | $\sigma_e^2 + 3\sigma_{GP/AE}^2 + 4\sigma_{SE/G/A}^2 + 12\sigma_{EG/A}^2$ |
| SE/G/A | 20 | $\sigma_e^2 + 4\sigma_{SE/G/A}^2$ |

| | | |
|---|---|---|
| $P/E$ | 6 | $\sigma_e^2 + 3\sigma_{GP/AE}^2 + 30\sigma_{P/E}^2$ |
| $AP/E$ | 6 | $\sigma_e^2 + 3\sigma_{GP/AE}^2 + 15\sigma_{AP/E}^2$ |
| $GP/AE$ | 48 | $\sigma_e^2 + 3\sigma_{GP/AE}^2$ |
| $SP/GE/A$ | 120 | $\sigma_e^2 + 3\sigma_{GP/AE}^2 + 15\sigma_{AP/E}^2$ |

$F' = (MS_{AE} + MS_{GP/AE})/(MS_{AP/E} + MS_{EG/A})$. The numerator $df =$ $(MS_{AE} + MS_{GP/AE})^2 / (MS_{AE}^2 / 1 + MS_{GP/AE}^2 / 48)$ and the denominator $df =$ $(MS_{AP/E} + MS_{EG/A})^2 / (MS_{AP/E}^2 / 6 + MS_{EG/A}^2 / 8)$.

**16.9**
(a)

| SV | df | EMS | Error Term |
|---|---|---|---|
| $A$ | 1 | $\sigma_e^2 + 3\sigma_{G/AP/E}^2 + 15\sigma_{AP/E}^2 + 120\theta_A^2$ | $AP/E$ |
| $E$ | 1 | $\sigma_e^2 + 3\sigma_{G/AP/E}^2 + 30\sigma_{P/E}^2 + 120\theta_E^2$ | $P/E$ |
| $AE$ | 1 | $\sigma_e^2 + 3\sigma_{G/AP/E}^2 + 15\sigma_{AP/E}^2 + 60\theta_{AE}^2$ | $AP/E$ |
| $P/E$ | 6 | $\sigma_e^2 + 3\sigma_{G/AP/E}^2 + 30\sigma_{P/E}^2$ | $G/AP/E$ |
| $AP/E$ | 6 | $\sigma_e^2 + 3\sigma_{G/AP/E}^2 + 15\sigma_{AP/E}^2$ | $G/AP/E$ |
| $G/AP/E$ | 64 | $\sigma_e^2 + 3\sigma_{G/AP/E}^2$ | $S/G/AP/E$ |
| $S/G/AP/E$ | 160 | $\sigma_e^2$ | |

(b) This analysis is likely to involve less error variance and provide more powerful, and certainly simpler, tests of $A$, $E$, and $AE$, However, monkeys are expensive to purchase and maintain, and the design of Exercise 16.8 involves far fewer subjects.

**16.10**

| SV | df | SS | MS | F | p |
|---|---|---|---|---|---|
| Leader (L) | 2 | 52.300 | 26.150 | 3.21 | .08 |
| G/L | 12 | 97.800 | 8.150 | | |
| Gender(X) | 1 | .017 | .017 | .01 | .92 |
| LX | 2 | 4.433 | 2.217 | .65 | .54 |
| XG/L | 12 | 40.800 | 3.400 | | |
| S/XG/L | 30 | 105.500 | 3.517 | | |

L is tested against G/L; X and LX are tested against XG/L.

**16.11**

(a) The result of the pseudogroup ANOVA is

| SV | df | SS | MS | F | p |
|---|---|---|---|---|---|
| Leader (L) | 1 | 30.625 | 30.625 | 6.646 | .03 |
| G/L | 8 | 85.100 | 10.637 | | |
| S/G/L | 30 | 138.250 | 4.608 | | |

(b) Let $NL$ be the noleader group condition and $I$ be the individual (no group) condition. Following the model in Table 16.7, $F' = MS_L/[(1/2)(MS_{G/NL} + MS_{S/I})]$. $MS_{G/NL}$ is $n \times$ the variance of the group means in the noleader condition; it can be obtained by doing an ANOVA of those 20 scores with *groups* as the independent variable. $MS_{S/I}$ is the variance of scores in the nogroup condition. Therefore, $F' = 26.150/(1/2)(8.825+5.937) = 26.150/7.381 = 3.543$. The error $df = 7.381^2/[(1/2)^2(8.825^2/4 + 5.937^2/19] \approx 10$; $p = .09$. In this instance, the $p$ value is considerably lower in the pseudogroup analysis but this may reflect the particular random assignment of individuals to pseudogroups in this example.

**16.12** Let $X = sex$, $R = relation$, (related/unrelated), $V = valence$ (positive or negative), and $I = items$. Note that items are nested in the $RV$ combinations. Sources of variance, $df$, sums of squares, and mean squares, may be obtained from software that permits the user to indicate nesting and crossing. In the absence of such software, any program that analyzes a repeated measure design can be used, although certain sources must be pooled. The $SV$, $df$, $SS$, and $MS$ are:

| SV | df | SS | MS |
|---|---|---|---|
| X | 1 | 8.533 | 8.533 |
| S/X | 22 | 443.333 | 20.152 |
| R | 1 | .533 | .533 |
| V | 1 | 38.533 | 38.533 |
| RV | 1 | 86.700 | 86.700 |
| I/RV | 22 | 28.200 | 1.282 |
| XR | 1 | 4.033 | 4.033 |
| XV | 1 | 1.200 | 1.200 |
| XRV | 1 | .033 | .033 |
| XI/RV | 16 | 21.034 | 1.315 |
| SR/X | 22 | 38.133 | 1.733 |
| SV/X | 22 | 42.567 | 1.935 |
| SRV/X | 22 | 70.767 | 3.217 |
| SI/RVX | 352 | 704.366 | 2.001 |

If the data were analyzed as if the same five items were present in each $RV$ cell, the three terms above involving $I$ can be obtained by pooling terms. $I/RV$ is the pool of $I$, $IR$, $IV$, and $IRV$; $I/RV$ is the pool of $XI$, $XIR$, $XIV$, and $XIRV$; and $SI/RVX$ is the pool of $SI/X$, $SIR/X$, $SIV/X$, and $SIRV/X$. The last four terms are error terms in the "non-nested" ANOVA for $I$, $IR$, $IV$, and $IRV$, respectively.

To test $V$, $F' = (MS_V + MS_{SI/RVX})/(MS_{I/RV} + MS_{SV/X}) = 12.6$. Following the general form of Equation 13.22, and rounding the result, the $df$s = 1 and 42, and $p$ = .001. To test $R \times V$, $F' = (MS_{RV} + MS_{SI/RVX})/(MS_{I/RV} + MS_{SRV/X}) = 19.716$, and the $df$ = 1 and 37; $p$ = .000. The means for the $RV$ cells are

|  | Related | Unrelated |
|---|---|---|
| Positive | 3.75 | 4.67 |
| Negative | 5.17 | 4.38 |

The negative prime results in a more negative average rating, but this is entirely due to the condition in which the prime is related to the target. This accounts for the valence main effect and the interaction of *valence* and *relation*. No other terms approach significance.

## Chapter 17

**17.1**

(a) $SS_A = 79.5$, $SS_C = 22.5$, $SS_S = 56$, and $SS_{res} = 29$.

(b) The estimates of the $S \times C$ effects and the scores after adjustment are

| | est$(\eta\gamma)_{ij}$ | | | | $Y_{ijk}$ - est$(\eta\gamma)_{ij}$ | | | |
|---|---|---|---|---|---|---|---|---|
| | $C_1$ | $C_2$ | $C_3$ | $C_4$ | $C_1$ | $C_2$ | $C_3$ | $C_4$ |
| $S_1$ | 3.500 | -5.000 | 2.250 | -0.750 | 21.500 | 21.000 | 21.750 | 18.750 |
| $S_2$ | 2.500 | 3.000 | -3.750 | -1.750 | 16.500 | 16.000 | 16.750 | 13.750 |
| $S_3$ | -4.500 | 1.000 | 2.250 | 1.250 | 17.500 | 17.000 | 17.750 | 14.750 |
| $S_4$ | -1.500 | 1.000 | -0.750 | 1.250 | 18.500 | 18.000 | 18.750 | 15.750 |

$SS_S$ and $SS_C$ are unchanged; however, $SS_A$ and $SS_{res}$ now both equal zero. The $S \times C$ interaction accounts for 9 *df*. Because the Latin square is not a fully factorial design, these same 9 *df* account for the $A$ variability (3 *df*) and the residual error variability (6 *df*). Picture a circle representing $S \times C$ variability. Part of the circle also represents the $A$ variability; the rest is the residual (error) variability. If there is an interaction of two variables in the population, effects attributed to the third variable may be due to the interaction. In contrast, if we had a fully factorial $S \times C \times A$ design, removal of any effect, main or interaction, would produce no change in the remaining sums of squares.

**17.2**

(a) The estimate of the repeated measures error term, $MS_{SA}$, follows from Equation 17.2; est$MS_{SA} = aMS_{res} + (MS_C - MS_{res})/a = 15.085$.

(b) From Equation 17.3, $RE_{LS \ to \ RM} = \left[ \dfrac{df_{LS} + 1}{df_{LS} + 3} \right] \cdot \left[ \dfrac{df_{RM} + 3}{df_{RM} + 1} \right] \cdot \left[ \dfrac{MS_{SA}}{MS_{res}} \right]$

$= \left[ \dfrac{6+1}{6+3} \right] \cdot \left[ \dfrac{9+3}{9+1} \right] \cdot \left[ \dfrac{19.999}{4.833} \right] = 3.86$. The repeated measures design is

estimated to need about 4 times as many subjects as the Latin square to attain the same power.

(c) The Latin square would be even more efficient because the added variance due to C would be part of the error tern in the repeated measures design but not in the Latin square design.

**17.3**

| SV | df | SS | MS | F | p |
|---|---|---|---|---|---|
| S | 3 | .593 | .198 | 47.4 | .000 |
| P | 3 | .232 | .077 | 18.6 | .002 |
| E | 1 | 1.323 | 1.323 | 197.6 | .000 |
| I | 1 | .903 | .903 | 144.4 | .000 |
| E x I | 1 | .023 | .023 | 3.6 | .107 |
| Residual | 6 | .025 | .004 | | |

**17.4**
**(a)**

| SV | df |
|---|---|
| Drug type (T) | 1 |
| R | 3 |
| TR | 3 |
| S/TR | 24 |
| Occasions (O) | 3 |
| Dosages (D) | 3 |
| TO | 3 |
| TD | 3 |
| B cells res | 6 |
| B cells res x T | 6 |
| W cells res | 72 |

*T*, *R*, and *TR* are tested against *S/TR*, and all other terms are tested against W cells res.

(b)

| SV | df |
| --- | --- |
| R | 7 |
| S/R | 8 |
| O | 7 |
| T | 1 |
| D | 3 |
| TD | 3 |
| B cells res | 42 |
| W cells res | 56 |

*R* is tested against *S/R*, and all other terms are tested against W cells res.

(c) Design (b) requires fewer subjects and involves a simpler analysis. On the other hand it may be impractical to run each subject on eight different occasions [as opposed to four in design (a)]. Of course, there is also some question as to whether either drug types or dosages should be within-subjects variables. The answer to that depends on how quickly the effect of a drug wears off and what recovery period is allowed between sessions.

**17.5** The simplest design is a 3 x 3 x 3 completely randomized design with three subjects in each of the 27 cells. This requires the least time from each subject, involves the fewest assumptions, runs no risk of carryover effects, and has 54 error *df*. However, it is also the least efficient design because the error term includes variance due to individual differences. A second possibility, would be to create a 3 x 3 Latin Square with one variable, perhaps (*W*) varied in a counterbalanced order. Twenty-seven subjects would be run in each row of the square (sequence of levels of *w*). These 27 individuals would be divided among the 9 combinations of the remaining two factors. The advantage of this design is that whichever factor is the within-subject factor, and its interactions, can be tested more efficiently than in the first, completely randomized, design. The disadvantage is the possibility of carryover effects, and the added assumptions (e.g., sphericity) involved in any within-subject design. A third possibility is a variation on the second in which two factors are within subjects, so that the basic design is a replicated Latin

Square. The advantage is that now two variables are efficiently tested. The possible disadvantages are those cited for the second design, and the added time for each participant. There are other possibilities, including Latin-Greco squares. The main disadvantage of this design is that it rests on the assumption of no interaction between the two within-subject variables.

**17.6**
(a) The digram balanced 6 x 6 square is

$$\begin{bmatrix} 1 & 6 & 2 & 5 & 3 & 4 \\ 2 & 1 & 3 & 6 & 4 & 5 \\ 3 & 2 & 4 & 1 & 5 & 6 \\ 4 & 3 & 5 & 2 & 6 & 1 \\ 5 & 4 & 6 & 3 & 1 & 2 \\ 6 & 5 & 1 & 4 & 2 & 3 \end{bmatrix}$$

(b) The digram balanced 5 x 5 square is

$$\begin{bmatrix} 1 & 5 & 2 & 4 & 3 \\ 2 & 1 & 3 & 5 & 4 \\ 3 & 2 & 4 & 1 & 5 \\ 4 & 3 & 5 & 2 & 1 \\ 5 & 4 & 1 & 3 & 2 \\ 3 & 4 & 2 & 5 & 1 \\ 4 & 5 & 3 & 1 & 2 \\ 5 & 1 & 4 & 2 & 3 \\ 1 & 2 & 5 & 3 & 4 \\ 2 & 3 & 1 & 4 & 5 \end{bmatrix}$$

**17.7** Note that $A$ and $C$ are tested against the within-cell residual ($WCR$).

| SV | df | SS | MS | F | p |
|----|----|----|----|----|----|
| R | 3 | 415.417 | 138.472 | 1.25 | .354 |
| S/R | 8 | 885.000 | 110.625 | | |
| A | 3 | 384.250 | 128.083 | 10.24 | .000 |
| C | 3 | 701.750 | 233.717 | 18.68 | .000 |
| BCR | 6 | 69.167 | 11.528 | .92 | .498 |
| WCR | 24 | 300.333 | 12.514 | | |

**17.8** $SS_S = SS_R + SS_{S/R} = 1300.417$. $SS_A$ is unchanged. The remaining variability would be attributed to the $SS_{SA}$. Therefore, the new ANOVA table is

| SV | df | SS | MS | F | p |
|----|----|----|----|----|----|
| Ss | 11 | 1300.417 | | | |
| A | 3 | 384.250 | 128.083 | 3.946 | .016 |
| SxA | 33 | 1071.250 | 32.462 | | |

The $F$ test of $A$ is negatively biased because considerable variability due to $C$ has been pooled with other terms to form a new error term. In this case, the result is still quite significant; however, if the original, correct, ANOVA had yielded a $p$ value of only slightly less than .05, we might have come to the a different, nonsignificant, result for the analysis performed here.

**17.9** The script means and the ANOVA table are:

| $S_1$ | $S_2$ | $S_3$ | $S_4$ |
|----|----|----|----|
| 2.321 | 1.909 | .766 | .617 |

| SV | df | SS | MS | F | $p_-$ |
|----|----|----|----|----|----|
| R | 3 | 10.375 | 3.458 | 1.80 | .188 |
| Ss/R | 16 | 30.825 | 1.927 | | |
| Script (S) | 3 | 42.462 | 14.154 | 53.82 | .000 |
| V | 1 | 40.527 | 40.527 | 154.95 | .000 |
| S/V | 2 | 1.895 | .948 | 3.60 | .035 |
| C | 3 | 5.953 | 1.984 | 7.54 | .000 |
| BCR | 6 | 1.783 | .297 | 1.13 | .359 |
| WCR | 48 | 12.606 | .263 | | |

We conclude that heart-beat change scores are affected by the script, and that this variance is primarily due to the difference between the negative and positive scripts; the change is greater for the two scripts with negative valence. All terms are tested against the within-cells residual (WCR) except R, which is tested against S/R. $SS_V$ can be calculated as a single df contrast:
$$SS_V = [(2.321 + 1.909) - (.766 + .617)]^2/(4/20) = 40.527. \quad SS_{S/V} = SS_S - SS_V.$$

**17.10**

| SV | df | SS | MS | F | p |
|----|----|----|----|----|----|
| R | 3 | 33.791 | 11.264 | 4.71 | .008 |
| H | 1 | .372 | .372 | .16 | .696 |
| RH | 3 | 25.673 | 8.558 | 3.58 | .025 |
| Ss/RH | 32 | 76.599 | 2.394 | | |
| Script (S) | 3 | 128.356 | 42.785 | 60.03 | .000 |
| V | 1 | 118.922 | 118.922 | 166.79 | .000 |
| S/V | 2 | 9.434 | 4.714 | 6.61 | .002 |
| HS | 3 | 5.075 | 1.692 | 2.37 | .075 |
| HV | 1 | .097 | .097 | .14 | .709 |

| | | | | | |
|---|---|---|---|---|---|
| *HS/V* | 2 | 4.978 | 2.489 | 3.49 | .034 |
| *C* | 3 | 4.544 | 1.515 | 2.13 | .102 |
| *BCR* | 6 | 2.195 | .366 | .51 | .799 |
| *H x BCR* | 6 | 1.009 | .168 | .24 | .962 |
| *WCR* | 96 | 68.426 | .713 | | |

*S/RH* is the error term for the terms above it in the table and *WCR* is the error term against which the remaining terms are tested.

## Chapter 18

**18.1** (b) $r = .620$;  (c) $r^2 = .385$; (d) $r^2 = .385$.

### 18.2

(a) The patients in the VA hospital may constitute a restricted sample.  For example, the anxiety scores may tend to be quite high.  The sample should not be used to make inferences about the general population.

(b) The correlation between height and weight for the mixed group would be expected to be lower than .60.  If the differences in mean height and weight for Martians and Jovians were great enough, the correlation might even be negative.

(c) This is a classic case of inferring causation from correlation.  People who graduate from college may indeed make more money, but it is not obvious how much of their financial success can be attributed directly to graduating from college.  Graduates may be smarter and more motivated and organized than nongraduates and therefore more successful.  Graduates  may also be more likely to come from wealthier and more stable families that are more willing to assist them.

**18.3** The reduction in the correlation between age and TC for older and younger women occurs because of the reduction in the variability of age – see the discussion in Section 18.2.2.  The corresponding figures for men are $r = .062$ for the overall correlation of age and TC in the sample;  $r = .008$ for men under 50 and $r = -.035$ for men 50 and over.

### 18.4

(a) For group 1, the correlation is $b_1 s_X / s_Y = (1)(10)/(20) = .50$; for group 2, the correlation is .80.

(b) The correlation between the transformed variables is still .70.

**18.5**

(a) Because $r_{XY} = r_{X'Y'}\sqrt{r_{XX}}\sqrt{r_{YY}}$ (Equation 18.7), the largest correlation that we could find would be $r_{XY} = \sqrt{r_{XX}}\sqrt{r_{YY}} = \sqrt{.64}\sqrt{.81} = .72$ .

(b) The estimated correlation if we "correct for attenuation" due to low reliability is $.40/\sqrt{r_{XX}}\sqrt{r_{YY}} = .40/\sqrt{.64}\sqrt{.81} = .40/.72 = .56$ .

(c) To test the correlation, we use the "uncorrected" correlation of .40. The test statistic is $t = \dfrac{r}{\sqrt{\dfrac{1-r^2}{N-2}}} = \dfrac{.40}{\sqrt{\dfrac{1-.16}{38}}} = 2.690$, $p < .05$. The correlation is significant.

**18.6** Considering only cases for men having data for both TC and age, we find the following means – for agegrp 1, mean TC =219.773, mean age = 33.796; for agegrp 2, mean TC =217.757, mean age = 44.685; for agegrp 3, mean TC = 228.585, mean age = 54.429; and for agegrp 4, mean TC =221.102, mean age = 64.770. The correlation between mean TC and mean age is .394.

**18.7**

(a) Using Equation 18.9, $t\,(17) = -1.30$, so we cannot reject $H_0$.

(b) $z = \dfrac{Z_r - Z_{hyp}}{\sqrt{\dfrac{1}{N-3}}} = \dfrac{-.310 - 0}{\sqrt{\dfrac{1}{16}}} = -1.24$, so again we cannot reject $H_0$.

(c) No, even if the correlation had been significant, we could not conclude that studying interferes with test performance. More likely, students having difficulty may study more, but still perform more poorly.

(d) The .95 CI for $Z_\rho$ is given by

$Z_r \pm z_{.025}\sqrt{\dfrac{1}{N-3}} = -.310 \pm (1.96)(1/4) = -.80, .18$. Translating back to correlations, the .95 CI for $\rho$ extends from -.66 to +.18. The .50 CI for $Z_\rho$ is -.310 $\pm(.675)(1/4) = -.48$ to -.14. Translating back to correlations, the interval extends from -.44 to -.14.

(e) Using GPOWER, the post hoc power is approximately .25.

(f) To calculate directly as in Table 18.1, recall from part (b) that the test statistic was $z = -1.24$. The lower critical value therefore has a z-score of -1.96 -(-1.24) = -.72 with respect to the sampling distribution given $\rho = .30$, so that the power is $p\,(z < -.72) = .24$, about the same value given by GPOWER.

(g) Using GPOWER, we need $N = 82$ to get power of .80.

(h) Using the procedure illustrated in Table 18.2, we get $N = 85$.

**18.8**

(a) $H_0: \rho_1 = \rho_2$ ; $H_1: \rho_1 \neq \rho_2$ ; $\alpha = .05$ . Use the test statistic

$$z = \frac{Z_{r_1} - Z_{r_2}}{\sqrt{\dfrac{1}{N_1 - 3} + \dfrac{1}{N_2 - 3}}} = \frac{.365 - .224}{\sqrt{2/197}} = 1.39, \text{ so we cannot reject } H_0 .$$

(b) Using the procedure of Table 18.3, the post hoc power is $p(z > .57) = .28$.

(c) The $N$ needed to get power of .80 is $N = 2\left[\dfrac{1.96 + .84}{.365 - .224}\right]^2 + 3 \approx 792$.

**18.9**

(a) $z = (Z_{.45} - 0)/\sqrt{1/49} = 3.395$, so we can reject $H_0$ .

(b) $z = \dfrac{Z_{.60} - Z_{.45}}{\sqrt{\dfrac{1}{N_1 - 3} + \dfrac{1}{N_2 - 3}}} = \dfrac{.693 - .485}{\sqrt{\dfrac{1}{49} + \dfrac{1}{64}}} = 1.193$, we cannot reject $H_0$

(c) $z = \dfrac{Z_{.45} - Z_{.20}}{\sqrt{\dfrac{1}{49} + \dfrac{1}{64}}} = 1.486$; the correlations do not differ significantly.

(d) The .95 CI for $Z_\rho$ is $.485 \pm (1.96)(1/7) = .205, .765$; the interval for $\rho$ is $.202, .644$.

**18.10**

(a) $\chi_3^2 = 36[.310^2 + .549^2 + .203^2] = 15.794$. The critical value of $\chi_{3,.05}^2$ is 7.815. So we can reject the hypothesis that the off-diagonal correlations are all 0 in the population.

(b) Using Steiger's MULTICORR program, we obtain $\chi_1^2 = 1.1791$, $p = .2270$. We can't reject the null hypothesis.

(c) The partial correlation is $r_{AV|Q} = \dfrac{.50 - (.30)(.20)}{\sqrt{(1 - .30^2)(1 - .20^2)}} = .47$.

The test statistic is $t = \dfrac{r}{\sqrt{\dfrac{1 - r^2}{N - 3}}} = 3.20$ with 36 $df$. The partial correlation is significant.

**18.11**

(a) Using Steiger's MULTICORR program, we find $\chi_1^2 = 34.007$, $p = .000$. The correlation between masculinity and femininity is significantly different at times 1 and 2.

(b) Again using the program, we find $\chi_1^2 = 0.798$, $p = .375$. We cannot reject the hypothesis that the correlation between V1 and M1 is the same as the correlation between V1 and F1 in the population.

**18.12**

(a) $\phi = -.36$; (b) $\phi_{max} = .56$, $\phi_{min} = -.80$; (c) To obtain $\phi = 0$, we need independence, so that, $p$(pass item 1 and pass item 2) $= p$(pass 1)$p$(pass 2) $= .4 \times .7 = .28$. The frequency of the pass-pass cell would then be $Np = 28$. The other frequencies may be filled in so as to preserve the marginals.

**18.13** $r_{X,XT} = \dfrac{1}{N-1}\sum z_X z_{T-X} = \dfrac{1}{N-1}\sum \left(\dfrac{X - \bar{X}}{s_X}\right)\left(\dfrac{T - X - \overline{T - X}}{s_{T-X}}\right)$

$$= \frac{1}{N-1}\sum \frac{(X - \bar{X})\left[(T - \bar{T}) - (X - \bar{X})\right]}{s_X \sqrt{\dfrac{1}{N-1}\sum\left[(T - \bar{T}) - (X - \bar{X})\right]^2}} = \frac{r_{XT} s_X s_T - s_X^2}{s_X \sqrt{s_T^2 + s_X^2 - 2r_{XT} s_X s_T}}$$

$$= \frac{r_{XT} s_T - s_X}{\sqrt{s_X^2 + s_T^2 - 2r_{XT} s_X s_T}}.$$

**18.14** Substituting into the equation, we obtain $r_{\text{pre,change}} = -.39$.

**18.15** There are several ways of doing the problem. We could start by expressing $r_{XT} = r_{X(X+Y)}$ in terms of $r_{XY}$:

$$r_{X(X+Y)} = \frac{1}{N-1}\sum z_X z_{X+Y} = \frac{1}{N-1}\sum\left(\frac{X - \bar{X}}{s_X}\right)\left(\frac{X + Y - \overline{X+Y}}{s_{X+Y}}\right)$$

$$= \frac{1}{N-1}\sum\left(\frac{X - \bar{X}}{s_X}\right)\left(\frac{(X - \bar{X}) + (Y - \bar{Y})}{\sqrt{s_X^2 + s_Y^2 + 2r_{XY} s_X s_Y}}\right) = \frac{s_X^2 + r_{XY} s_X s_Y}{s_X \sqrt{s_X^2 + s_Y^2 2r_{XT} s_X s_Y}}$$

$$= \frac{s_X + r_{XY} s_Y}{\sqrt{s_X^2 + s_Y^2 + 2r_{XY} s_X s_Y}}. \quad \text{Now, if } s_X = s_Y = s \text{, then}$$

$$r_{X(X+Y)} = \frac{s^2 + r_{XY} s^2}{s\sqrt{s^2 + s^2 + 2r_{XY} s^2}} = \sqrt{\frac{1 + r_{XY}}{2}} = r_{XT} = .70.$$

Squaring both sides, $\dfrac{1 + r_{XY}}{2} = .49$, so that $r_{XY} = -.02$. Here, the two parts of the test are not correlated. The reason $r_{XT}$ is high is because $X$ is part of $T$.

## Chapter 19

**19.1**

(a) The strategy is not a good one. Given inconsistent behavior, very bad performance is likely to be followed by better performance, and exceptionally good performance may well be followed by performance that is not as good, whether or not feedback is given.

(b) The improvement cannot be explained as regression toward the mean.

Regression toward the mean by itself would not account for above average performance by the group that received tutoring.

(c) Not necessarily. Regression toward the mean complicates matching. Suppose that good spellers on the average have much higher IQ's than poor spellers. If we were to form a mixed group of good and poor spellers matched for IQ on the basis of a single test, then the mixed group might largely consist of more intelligent good spellers who just happened to perform poorly on the IQ test and less intelligent poor spellers who performed well on the test. If they were to be given a second IQ test, the two groups might regress to different means.

**19.2**

(a) The best prediction is 73 inches.

(b) The best prediction for the height of the father is 71.5. Again there is regression toward the mean.

(c) Even though there is regression toward the mean for the predicted score whenever the absolute value of the correlation is less than one, this does not imply the distribution of height will become less variable over time. Whenever there is less than perfect prediction, the actual scores can be thought of as being distributed around the prediction. If these distributions are summed, the original distribution should be reproduced.

**19.3**

(a) The scatterplot indicates that there is a strong curvilinear relation between the two variables:

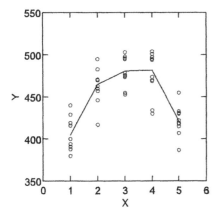

Results of the regression of $Y$ on $X$:

```
Dep Var: Y   N: 50   Multiple R: 0.178   Squared multiple R: 0.032
Adjusted squared multiple R: 0.012   Standard error of estimate: 36.309
```

| Effect | Coefficient | Std Error | Std Coef | Tolerance | t | P(2 Tail) |
|---|---|---|---|---|---|---|
| CONSTANT | 435.880 | 12.042 | 0.000 | . | 36.196 | 0.000 |
| X | 4.560 | 3.631 | 0.178 | 1.000 | 1.256 | 0.215 |

```
Analysis of Variance
Source      Sum-of-Squares   df      Mean-Square  F-ratio   P
Regression    2079.360       1        2079.360    1.577     0.215
Residual     63280.960      48        1318.353
  Durbin-Watson D Statistic          0.677
First Order Autocorrelation          0.633
```

Results of an ANOVA with $X$ as the factor :

```
Analysis of Variance
Source    Sum-of-Squares  df      Mean-Square   F-ratio    P
X           45790.520      4       11447.630     26.323    0.000
Error       19569.800     45         434.884
```

To test whether there is also a significant linear effect, we note that from the regression, $SS_{linear} = MS_{linear} = 2079.360$. To test for linearity, form the ratio $F = MS_{linear}/MS_{error}$, where $MS_{error}$ is obtained from the ANOVA; $F(1, 45) = 2079.360/434.884 = 4.781$; there is a significant linear effect, $p < .05$. The best-fitting regression equation is $\hat{Y} = 435.88 + 4.56X$; this predicts values of 440.44, 445.00, 449.56, 454.12, and 458.68 for $X = 1$-5, respectively.

(b) The plot of residuals ($Y$ against $\hat{Y}$) is curvilinear, indicating that a curvilinear component in the relation between $Y$ and $X$ has not been accounted for by the regression.

(c) We can now fill in the table:

| SV | df | SS | MS | F | p |
|---|---|---|---|---|---|
| Between | 4 | 45790.52 | 11447.63 | 26.32 | .000 |
| Linearity | 1 | 2079.36 | 2079.36 | 4.78 | < .05 |
| Lack of Fit (nonlinearity) | 3 | 43711.16 | 14570.39 | 33.50 | < .001 |
| Pure Error | 45 | 19569.80 | 434.88 | | |

(d) We can create a new variable XSQ $= X*X$ and regress $Y$ on X and XSQ. This yields the following output:

```
Dep Var: Y   N: 50   Multiple R: 0.833   Squared multiple R: 0.693
Adjusted squared multiple R:0.680 Standard error of estimate: 20.657
```

```
Effect   Coefficient Std Error Std Coef  Tolerance     t      P(2 Tail)
CONSTANT   312.880     14.010    0.000      .        22.332   0.000
X          109.989     10.677    4.302     0.037     10.302   0.000
XSQ        -17.571      1.746   -4.203     0.037    -10.065   0.000
```

```
Analysis of Variance
Source      Sum-of-Squares   df      Mean-Square  F-ratio   P
Regression   45305.074       2       22652.537    53.087    0.000
Residual     20055.246      47         426.707
```

```
Durbin-Watson D Statistic          1.793
First Order Autocorrelation        0.101
```

The regression equation is $\hat{Y} = 312.88 + 109.99X + 17.571X^2$. The quadratic component is highly significant. $R^2$ is now .69 as opposed to .03 in the original regression. The residual plot no longer suggests any obvious nonlinearity.

### 19.4

(a) For the 211 females having data on both cholesterol level and age, the regression of cholesterol level ($Y$) on age ($X$) yields $\hat{Y} = 131.870 + 1.712X$. The standard error of estimate is 34.219. $SE(b_0) = 10.053$ and $SE(b_1) = 0.202$.

(b) For 30-year-old women $\hat{\mu}_{Y.X=30} = \hat{Y} = 131.870 + 1.712(30) = 183.23$. For 50-year-olds, $\hat{\mu}_{Y.X=50} = 217.47$. The .95 CI's for the conditional means my be found using $\hat{\mu}_{Y.X} \pm t_{.025,209} SE(\hat{\mu}_{Y.X})$, where $SE(\hat{\mu}_{Y.X}) = s_e \sqrt{\dfrac{1}{N} + \dfrac{(X - \bar{X})^2}{SS_X}}$.

We can find $SS_X$ several ways – one way is to note that $SE(b_1) = s_e / \sqrt{SS_X} = 0.202$, so that $SS_X = (s_e/.202)^2 = (34.219/0.202)^2 = 28696.70$. Now we need the mean $X$ score for the 211 women having data on cholesterol and age; this is 48.398. Substituting, we find that $SE(\hat{\mu}_{Y.X})$ is 4.400 for 30-year-olds and 2.378 for 50-year-olds. Given that $t_{.025,209} = 1.971$, the .95 CI for 30-year-old women is 174.56, 191.90; for 50-year-olds it is 212.78, 222.16.

(c) The interval is narrower (and hence the estimate is more likely to be closer to the population parameter) for 50-year-old women because 50 is closer to the mean age (43.898) than 30, and hence the standard error is smaller.

(d) Now the appropriate standard error is

$$s_e \sqrt{1 + \frac{1}{N} + \frac{(X - \bar{X})^2}{SS_X}} = (34.218)\sqrt{1 + \frac{1}{211} + \frac{(30 - 43.898)^2}{28696.70}} = 34.415$$

and the .95 CI extends from 115.40 to 251.06.

### 19.5

(a) Using Equation 18.13, $z = 3.38$, so there is a higher correlation between salary and years of service for men than for women – that is, there is a stronger linear relation for men than for women.

(b) Using $b_1 = rs_Y / s_X$, we find there is a slope of .599 for men and .799 for women. Each additional year of service corresponds to about an additional \$600 (.599 x \$1000) for men and about \$800 for women. We can test whether the slope difference is significant using the test statistic

$$t = \frac{b_M - b_W}{SE(b_M - b_W)} = \frac{b_M - b_W}{s_e \sqrt{\dfrac{1}{SS_{X_M}} + \dfrac{1}{SS_{X_W}}}}$$

where $s_e^2 = \dfrac{SS_{residual}}{N_M + N_W - 4} = \dfrac{(1-r_M^2)SS_{Y_M} + (1-r_W^2)SS_{Y_W}}{3996} = \dfrac{575855.86 + 545806.95}{3996}$

$= 280.70$, so that $s_e = 16.75$. Substituting, we find $t\,(3996) = -2.38$, $p < .02$. The salary increment per year for women is significantly greater than that for men. So here we have a situation in which the correlation is significantly larger for men than women but the slope is significantly higher for women than for men. The reason for this apparent paradox is that the men have greater variability in their years of service.

**19.6** (a) The results of the regression of $Y$ on Dosage ($D$) are as follows:

```
Dep Var: Y   N: 20   Multiple R: 0.758   Squared multiple R: 0.574
Adjusted squared multiple R: 0.551   Standard error of estimate: 12.707
```

| Effect | Coefficient | Std Error | Std Coef | Tolerance | t | P(2 Tail) |
|---|---|---|---|---|---|---|
| CONSTANT | 6.900 | 6.960 | 0.000 | . | 0.991 | 0.335 |
| DOSAGE | 1.252 | 0.254 | 0.758 | 1.000 | 4.926 | 0.000 |

Analysis of Variance

| Source | Sum-of-Squares | df | Mean-Square | F-ratio | P |
|---|---|---|---|---|---|
| Regression | 3918.760 | 1 | 3918.760 | 24.269 | 0.000 |
| Residual | 2906.440 | 18 | 161.469 | | |

The regression equation is $\hat{Y} = 6.90 + 1.25D$. The slope is significantly different from 0, $t\,(18) = 4.93$, p $= .000$.

If we perform an ANOVA, the output is

```
Dep Var: Y   N: 20   Multiple R: 0.858   Squared multiple R: 0.736
Analysis of Variance
```

| Source | Sum-of-Squares | df | Mean-Square | F-ratio | P |
|---|---|---|---|---|---|
| DOSAGE | 5022.000 | 3 | 1674.000 | 14.854 | 0.000 |
| Error | 1803.200 | 16 | 112.700 | | |

The $D$ main effect is significant. The hypothesis tested is that the population means for the different levels of dosage are identical.

**19.7** (a) Regressing FINAL on PRETEST produces the following output:

```
Dep Var: FINAL   N: 18   Multiple R: 0.725   Squared multiple R: 0.526
Adjusted squared multiple R: 0.496   Standard error of estimate: 10.638
```

| Effect | Coefficient | Std Error | Std Coef | Tolerance | t | P(2 Tail) |
|---|---|---|---|---|---|---|
| CONSTANT | -36.083 | 27.295 | 0.000 | . | -1.322 | 0.205 |
| PRETEST | 3.546 | 0.842 | 0.725 | 1.000 | 4.212 | 0.001 |

Analysis of Variance

| Source | Sum-of-Squares | df | Mean-Square | F-ratio | P |
|---|---|---|---|---|---|
| Regression | 2007.497 | 1 | 2007.497 | 17.738 | 0.001 |
| Residual | 1810.780 | 16 | 113.174 | | |

(b) The regression equation is $\hat{\text{FINAL}} = -36.08 + 3.55\text{PRETEST}$. The standard error of estimate is 10.64 and the standard errors of $b_0$ and $b_1$ are 27.295 and

0.842, respectively.

(c) Using the regression equation, estimates of the conditional means of the population of FINAL scores for PRETEST = 24 and 37 are 49.02 and 95.12, respectively. To find the confidence intervals for the conditional means, we need the standard errors for the predicted final scores,

$$SE(\stackrel{\wedge}{\text{FINAL}}) = s_e \sqrt{h_{jj}} = s_e \sqrt{\frac{1}{N} + \frac{(X - \bar{X})^2}{SS_X}} \text{ . For PRETEST=24, } N = 18;$$

$(X - \bar{X})^2 = (24 - 32.278)^2; SS_X = (N-1)s_X^2 = 159.60$ . So,

$$SE(\stackrel{\wedge}{\text{FINAL}}) = (10.638)\sqrt{\frac{1}{18} + \frac{(24 - 32.278)^2}{159.60}} = 7.408 \text{ . Similarly, the } SE \text{ for}$$

PRETEST=37 is 4.708. The .95 CI for the conditional mean of FINAL scores at PRETEST=24 is given by $49.02 \pm t_{16,.025} SE = 49.02 \pm (2.12)(7.408) = 49.02 \pm 15.71$. Similarly, the .95 CI at PRETEST = 37 is given by $95.12 \pm 9.98$.

(d) The estimate at PRETEST=37 is likely to be more accurate. Because it is closer to the mean of the PRETEST scores, it has a smaller leverage and therefor a smaller standard error.

(e) To find the .95 CI for the FINAL score of a single student with PRETEST score = 24, we need the appropriate standard error given by

$$s_e\sqrt{1 + \frac{1}{N} + \frac{(X - \bar{X})^2}{SS_X}} = (10.638)\sqrt{1 + \frac{1}{18} + \frac{(24 - 32.378)^2}{159.60}} = 12.96 \text{ , so the CI is}$$

$49.02 \pm 27.48$.

### 19.8

(a) No, they are testing different hypotheses. Anne is testing whether there are any effects of array size. Her null hypothesis is $\mu_1 = \mu_2 = \mu_3 = \mu_4$. On the other hand, Reg is testing the hypothesis $H_0: \beta_1 = 0$. He is concerned whether response time varies linearly with array size. If Reg's hypothesis is false, then so is Anne's, but the reverse is not necessarily true.

(b) An ANOVA can be performed using $F = MS_A / MS_{S/A}$, where $SS_A = n\sum_j(\bar{Y}_{.j} = \bar{Y}..)^2 = 10[(-40)^2 + 0^2 + 20^2 + 20^2] = 24,000$, and $MS_A = 24,000/3$ $= 8000$. Also, $MS_{S/A} = \sum s_j^2 / a = \frac{360 + 315 + 324 + 333}{4} = 333$ ; therefore, $F(3, 36) = 8000/333 = 24.02$. This is highly significant. Also, note $SS_{S/A} = (36)(333) = 11,988$.

To test $H_0: \beta_1 = 0$, use $t = b_1 / SE(b_1)$, where $SE(b_1) = s_e / \sqrt{SS_X}$ ;

$$b_1 = \frac{\sum\sum(X_j - \bar{X})(Y_{ij} - \bar{Y}..)}{SS_X} = \frac{(10)[(-3)(-40) + (-1)(0) + (1)20 + (3)(20)}{(10)[(-3)^2 + (-1)^2 + 1^2 + 3^2]} = 10$$

To find $s_e$, we must first find $SS_{residual} = SS_Y - SS_{regression}$. $SS_{regression} = b_1^2 SS_X =$
$(100)(200) = 20,000$. $SS_Y = SS_{total} = SS_A + SS_{S/A} = 24,000 + 11,988 = 35,988$.
Therefore, $SS_{residual} = 35,988 - 20,000 = 15,988$, so that $s_e = \sqrt{SS_{residual}/(N-2)}$
$= \sqrt{15,988/38} = 20.51$ and $SE(b_1) = 20.51/\sqrt{200} = 1.45$. The null hypothesis is
tested by $t = b_1/SE(b_1) = 10/1.45 = 6.90$. The result is highly significant.

(c) $SS_A$ must always be at least as large as $SS_{regression}$ and will be larger
unless the means for each of the conditions fall exactly on the regression line (see
Chapter 21 for a more complete explanation.

(d) $SS_{nonlinearity} = SS_{residual} - SS_{S/A} = 15,988 - 11,988$. Therefore, $SS_{nonlinearity} =$
$4,000$. The hypothesis that there is no nonlinearity is tested by $F(2, 36) =$
$MS_{nonlinearity}/MS_{S/A} = (4000/2)/333 = 6.01$. The hypothesis can be rejected at $p < .01$.

**19.9** The null hypothesis that the population slopes are equal can be tested
using the test statistic $t = \dfrac{b_{1_M} - b_{1_W}}{SE(b_{1_M} - b_{1_W})}$ where $SE(b_{1_M} - b_{1_W})$ can be estimated by
$s_e\sqrt{\dfrac{1}{SS_{X_M}} + \dfrac{1}{SS_{X_W}}}$ (see Table 19.8). In general, $s_e^2$ is the weighted average of
$s_{e_M}^2$ and $s_{e_W}^2$, here 194.55, so $s_e = 13.95$. Therefore, we test the null hypothesis
using $t = \dfrac{30.0 - 20.0}{13.95\sqrt{\dfrac{1}{200} + \dfrac{1}{200}}} = 7.17$ with $N_M - 2 + N_W - 2 = 76$ $df$. We can reject the
null hypothesis.

**19.10**

(a) The individual slopes are simple treated as scores and subjected to an
independent-groups t test yielding the following output. The .95 CI extends from
.91 to 10.49.

```
Data for the following results were selected according to:
    (MATERIALS= letters)

Two-sample t test on Y grouped by SEX
```

| Group | N | Mean | SD |
|---|---|---|---|
| 1 | 10 | 27.400 | 6.132 |
| 2 | 10 | 21.700 | 3.802 |

```
    Separate Variance t =     2.498 df =  15.0     Prob = 0.025
    Difference in Means =     5.700   95.00% CI = 0.838 to 10.562

    Pooled Variance t =       2.498 df =   18      Prob = 0.022
    Difference in Means =     5.700   95.00% CI = 0.907 to 10.493
```

(b) The results of the *t* test for digits are as follows:
```
Data for the following results were selected according to:
    (MATERIALS= letters)
```

Two-sample t test on Y grouped by SEX

| Group | N | Mean | SD |
|-------|-----|--------|-------|
| 1 | 10 | 26.100 | 6.724 |
| 2 | 10 | 21.700 | 3.592 |

Separate Variance t = 1.825 df = 13.7 Prob = 0.090
Difference in Means = 4.400 95.00% CI = -0.779 to 9.579

Pooled Variance t = 1.825 df = 18 Prob = 0.085
Difference in Means = 4.400 95.00% CI = -0.665 to 9.465

The slopes are not significantly different for males and females.

(c) If we perform an ANOVA on the data, the results are:

Dep Var: Y   N: 40   Multiple R: 0.458   Squared multiple R: 0.210

Analysis of Variance

| Source | Sum-of-Squares | df | Mean-Square | F-ratio | P |
|--------|---------------|-----|-------------|---------|-------|
| MATERIALS | 4.225 | 1 | 4.225 | 0.153 | 0.698 |
| SEX | 255.025 | 1 | 255.025 | 9.260 | 0.004 |
| MATERIALS*SEX | 4.225 | 1 | 4.225 | 0.153 | 0.698 |
| Error | 991.500 | 36 | 27.542 | | |

There are no significant Sex x Materials interaction or materials main effect. There is a significant sex main effect, $F(1, 36) = 9.260, p = .004$.

**19.11** If we regress TC on age for males we get the following results:

Dep Var: TC   N: 220   Multiple R: 0.062   Squared multiple R: 0.004
Adjusted squared multiple R: 0.000   Standard error of estimate: 38.286

| Effect | Coefficient | Std Error | Std Coef | Tolerance | t | P(2 Tail) |
|--------|-------------|-----------|----------|-----------|--------|-----------|
| CONSTANT | 211.906 | 11.249 | 0.000 | . | 18.837 | 0.000 |
| AGE | 0.198 | 0.218 | 0.062 | 1.000 | 0.912 | 0.363 |

Analysis of Variance

| Source | Sum-of-Squares | df | Mean-Square | F-ratio | P |
|--------|---------------|-----|-------------|---------|-------|
| Regression | 1218.673 | 1 | 1218.673 | 0.831 | 0.363 |
| Residual | 319540.841 | 218 | 1465.784 | | |

*** WARNING ***
Case          409 is an outlier          (Studentized Residual = 4.362)

Durbin-Watson D Statistic          2.011
First Order Autocorrelation       -0.018

The slope is not significant, $t(218) = 0.912, p = .363$. To check assumptions, we begin by noting that the residual plot doesn't show any obvious nonlinearity, although there is perhaps a tendency for the lower estimates to have larger residuals, indicating a possible problem with homogeneity of variance. If we use perform a weighted least squares regression, as described in Section 19.5.3, we get the results that are not greatly different from those obtained in the original analysis.

| Effect | Coefficient | Std Error | t | P(2 Tail) |
|---|---|---|---|---|
| CONSTANT | 212.89 | 12.27 | 17.36 | 0.000 |
| AGE | 0.180 | 0.22 | 0.81 | 0.42 |

We can check for nonlinearity by first conducting an ANOVA in which age is the independent variable. We obtain the following output:

```
Dep Var: TC   N: 220   Multiple R: 0.461   Squared multiple R: 0.212
Analysis of Variance
```

| Source | Sum-of-Squares | df | Mean-Square | F-ratio | P |
|---|---|---|---|---|---|
| AGE | 68150.767 | 46 | 1481.538 | 1.015 | 0.457 |
| Error | 252608.747 | 173 | 1460.166 | | |

$SS_{nonlinearity} = SS_{residual} - SS_{error} = 319,540.841 - 252,608.094 = 66,932.094$ with 45 $df$. So we test the null hypothesis of no nonlinearity using $F(45, 173) = MS_{nonlinearity}/MS_{error}$ $= (66932.094/45)/1460.166 = 1.019$, so there is no significant nonlinearity.

The Durbin-Watson test doesn't how any evidence of serial correlation; $D$ is close to 2 and $r$ is only -.018. SYSTAT identifies the case with ID=686 as an outlier, having an externally studentized residual of 4.362. This score comes from a 32-year-old male with very high cholesterol readings (all seasonal TC levels at least 358). But even this case is not extremely influential. The Cook's distance is .137 – much less than $F_{.50}(2,218) = .695$. Redoing the regression omitting this case results in a slope of 0.294, higher, but not greatly different from the original value.

There is some modest non-normality. Plots of the residual indicate somewhat heavy tails and a slight positive skew. The kurtosis = 1.485 and skewness = .616. Finally when we tried out several types of robust regression, the obtained results were not very different from those in the original regression. We therefore conclude that the assumption for the original least-squares regression were reasonably well satisfied.

## Chapter 20

**20.1**

(a)

```
Pearson correlation matrix
              TIME        NUMBER        DIFF
TIME         1.000
NUMBER       0.756       1.000
DIFF         0.339       0.000        1.000
```

| Number = | 2 | 2 | 4 | 4 | 6 | 6 | 8 | 8 |
|---|---|---|---|---|---|---|---|---|
| Diff = | 10 | 20 | 10 | 20 | 10 | 20 | 10 | 20 |

| Mean: | 493.33 | 532.00 | 545.67 | 583.33 | 584.33 | 646.67 | 625.67 | 670.00 |
|---|---|---|---|---|---|---|---|---|
| SD: | 10.50 | 35.68 | 62.96 | 40.53 | 30.37 | 44.74 | 70.54 | 40.04 |

(b) In the regression, the effects of both number and diff are significant, $t(21) = 6.192$, $p = .000$ and $t(21) = 2.775$, $p = .011$, respectively. If an ANOVA is conducted, we find significant main effects for both number and diff, $F(3,16) = 10.216$, $p = .001$ and $F(1,16) = 6.091$, $p = .025$, respectively. The results of the regression are not equivalent to that of an ANOVA. The regression treats the predictors as quantitative variables and tests whether the rate of change of time with one of the variables is different from 0 in the population, holding the other variable constant. In the ANOVA, the test of the number main effect addresses the question of whether the population means for the different levels of number are all the same.

(c) The regression yields estimates of 402.375, 22.825, and 4.575 for $\beta_0$, $\beta_1$, and $\beta_2$. The 95% confidence intervals are 335.99 - 468.76, 15.16 - 30.49, and 1.157 - 8.00. The 99% confidence intervals are 312.01 - 492.75, 12.39 - 33.27, and -0.09 - 9.25.

**20.2** (c) Here the regression coefficient of $X_1$ is the same (1.000) whether or not $X_2$ is included in the regression. This will be true in general only if $X_1$ and $X_2$ are uncorrelated. However, the standard error for the coefficients of $X_1$ does not stay the same in the two regressions.

**20.3**

(a) Regressing $Y$ on $X_1$ and $X_2$ yields the regression equation
$$\hat{Y} = -16.294 + 9.196X_1 + 9.941X_2 .$$

(b) The ANOVA table for the regression indicates that $SS_{\text{residual}} = 1030.745$ with 14 $df$. Next, to obtain $SS_{\text{pure error}}$, we perform an ANOVA, using $X_1$ and $X_2$ as factors. The error term of the ANOVA is $SS_{\text{pure error}} = 937.417$ with 10 $df$. Therefore, $SS_{\text{nonlinearity}} = 1030.745 - 937.417 = 93.328$ with 4 $df$. $F = MS_{\text{nonlinearity}}/MS_{\text{pure error}} = .249$, so there is no significant departure from linearity.

(c) In the initial regression, the coefficients of $X_1$ and $X_2$ both differ significantly from 0, $t(14) = 2.420$, $p = .030$ and $t(14) = 3.378$, $p = .005$, respectively. Therefore, both variables should be included.

**20.4** (a) .10; (b) .44; (c) .49; (d) .26

**20.5**

(a) For the data set, a standard ANOVA yields

| SV | df | SS | MS | F | p |
|---|---|---|---|---|---|
| Dosage (D) | 3 | 132.306 | 44.102 | 7.290 | .003 |
| Error (S/D) | 15 | 90.740 | 6.049 | | |

(b) A trend analysis yields

| SV | df | SS | MS | F | p |
|---|---|---|---|---|---|
| Dosage (D) | 3 | 132.306 | 44.102 | 7.290 | .003 |
| linear | 1 | 62.110 | 62.110 | 10.268 | < .01 |
| quadratic | 1 | 65.375 | 65.375 | 10.808 | < .01 |
| cubic | 1 | 4.821 | 4.821 | 0.797 | ns |
| Error (S/D) | 15 | 90.740 | 6.049 | | |

(c) The best-fitting quadratic equation is

$$\hat{Y} = -4.714 + 1.122X - 0.019X^2 .$$

## 20.6

(a) If we have three predictors and only four cases, unless one or more cases are completely redundant, the resulting regression equation must predict $Y$ perfectly for the four cases, so that $R_{Y.123} = 1$.

(b) and (c) The predicted scores are obtained using the equation obtained in part (a), $\hat{Y} = 0.770 + 1.182X_1 - 1.731X_2$. The predicted scores correlate perfectly with the actual scores for the first four cases but have a correlation of .244 with the remaining 11 cases. A regression equation that fits a sample need not fit the population well if the $N/p$ ratio is small.

## 20.7

(a) When individual bivariate regressions are performed, student background and parent background are both significant. However, the background measures are highly correlated and if both measures are included as predictors, the student background measure remains significant but the parental background measure does not. Also, although the teacher quality does not predict significantly by itself, it adds significantly to predictability if either student or parental background is included.

(b) Here, the stepwise regression enters student background on the first step and teacher quality on the second. Although these seem like useful predictors, the predictor variables that are first selected in a stepwise regression need not be the ones that are important in an explanatory model.

(c) There are no significant interactions between either of the background measures and either of the quality measures.

**20.8**

(a) $R^2 = .20$, so that $f^2 = .25$. Given 6 predictor variables and 40 cases, we can use GPOWER or Steiger's R2 program to determine the number of cases necessary to have power = .80 for the hypothesis test. GPOWER indicates that we need $N = 62$, whereas R2 indicates that we need $N = 64$.

**20.9** If including $X_4$ in the regression equation results in $R^2$ increasing from .21 to .27, then $\Delta R^2 = .06$ and $f^2 = .06/(1-.27) = .082$. To test whether the increment in predictability afforded by the addition of $X_4$ is significant, we form the partial-$F$ ratio

$$F = \frac{(R^2_{Y.1234} - R^2_{Y.123})/1}{(1 - R^2_{Y.1234})/(N - p - 1)} = \frac{.06}{(1-.27)/35} = 2.877$$

The obtained $F$ is less than the critical $F_{.05, 1, 35}$ of 4.121. To determine the $N$ necessary to have a power of .80 for the test of $X_4$, we try out several values for $N$, finding the noncentrality parameter and critical $F$ value for each, then use a noncentral $F$ calculator to determine the power. For $N = 101$, $N^* = 98$, the noncentrality parameter $\lambda$ is 8.036, and the critical value of $F$ (with 1 and 96 df) is 3.940. This yields estimated power of .801, approximately the desired value.

**20.10**

(a) Both externalization and criticism contribute significantly to attachment when the other is held constant. Each contributes significantly to the prediction of attachment over and above the contribution of the other predictor variable.

(b) Adding the new variable to the regression equation is an appropriate way of determining whether the joint effect (i.e., the interaction) of externalization and criticism contribute to the predictability of attachment over and above the contributions made by the two measures by themselves. Because of the correlation between the externalization and criticism measures and their product, it is important that the joint effect be evaluated in an equation that also contains the attachment and criticism measures. The interpretation of the coefficients of attachment and criticism are different in a regression equation that also contains the product term than in an equation in which the product term is omitted. In the latter equation, the interpretation of the coefficient of externalization is the rate of change of attachment with externalization if criticism is held constant. In the former equation, the interpretation of the coefficient is the rate of change of attachment with externalization if criticism is zero.

**20.11** The ANOVA table is as follows:

| SV | df | SS | MS | F | p |
|----|----|----|----|----|----|
| Grade | 7 | 10,040.060 | 1434.294 | 17.491 | .000 |
| Linear | 1 | 5,668.646 | 5,668.646 | 69.128 | .000 |
| Quadratic | 1 | 2,170.851 | 2,170.851 | 26.473 | .000 |
| Cubic | 1 | 684.947 | 684.947 | 8.351 | .002 |
| Other nonlin | 4 | 1,515.516 | 378.879 | 4.620 | .003 |
| Error | 185 | 15,170.367 | 82.002 | | |

There are highly significant linear, quadratic, and cubic components. There is a significant linear component – the best-fitting straight line has a slope other than 0. There are also significant tendencies to level off and possibly decrease as Grade increases, then to increase again. Some of the other bumps in the curve are probably due to other than chance variation because the remaining nonlinearity is significant.

**20.12**

(a) Yes, there is a relation between TC and BMI for males. For the 183 men with data on age, BMI and TC, the correlation between TC and BMI was $r = .215$, $p = .004$. When TC is regressed on both age and BMI, the coefficient of BMI is significant, $b = 2.139$, $t(180) = 2.948$, $p = .004$

(b) We don't have enough evidence to claim that there is a significant interaction. When the interaction between age and BMI was tested by creating a new variable that was the product of age and BMI, then regressing TC on age, BMI, and the product Age $x$ BMI, the coefficient of the product variable was almost as large as was found for women (-.120 as opposed to -.135 for women), however, it was not significant, $t(179) = -1.888$, $p = .061$.

# Chapter 21

**21.1**

(a) We can create a Gender variable with levels, say, 1 for males and 0 for females. If we correlate Gender with the dependent variable, we find $r(14) = -.509, p = .044$.

(b) We find that there is a significant effect of gender, $t(14) = 2.210, p = .044$. As we noted in Section 18.5.1, the test of the point-biserial correlation is equivalent to an independent-groups $t$ test.

(c) Because it has only two levels, we only need a single dummy variable to code for gender. If males and females were assigned values of 1 and -1, we would have effect coding, if the values were 1 and 0, we would have dummy coding.

(d) For effect coding we find the output

```
Dep Var: Y   N: 16   Multiple R: 0.509  Squared multiple R: 0.259
Adjusted squared multiple R: 0.206   Standard error of estimate: 5.769
```

| Effect | Coefficient | Std Error | Std Coef | Tolerance | t | P(2 Tail) |
|--------|-------------|-----------|----------|-----------|------|-----------|
| CONSTANT | 28.187 | 1.442 | 0.000 | . | 19.545 | 0.000 |
| EFFECT | -3.188 | 1.442 | -0.509 | 1.000 | -2.210 | 0.044 |

Analysis of Variance

| Source | Sum-of-Squares | df | Mean-Square | F-ratio | P |
|--------|----------------|----|-----|---------|---|
| Regression | 162.563 | 1 | 162.563 | 4.885 | 0.044 |
| Residual | 465.875 | 14 | 33.277 | | |

and for dummy coding, we obtain

```
Dep Var: Y   N: 16   Multiple R: 0.509   Squared multiple R: 0.259
Adjusted squared multiple R: 0.206   Standard error of estimate: 5.769
```

| Effect | Coefficient | Std Error | Std Coef | Tolerance | t | P(2 Tail) |
|--------|-------------|-----------|----------|-----------|------|-----------|
| CONSTANT | 31.375 | 2.040 | 0.000 | . | 15.384 | 0.000 |
| DUMMY | -6.375 | 2.884 | -0.509 | 1.000 | -2.210 | 0.044 |

Analysis of Variance

| Source | Sum-of-Squares | df | Mean-Square | F-ratio | P |
|--------|----------------|----|-----|---------|---|
| Regression | 162.563 | 1 | 162.563 | 4.885 | 0.044 |
| Residual | 465.875 | 14 | 33.277 | | |

Note that the ANOVA tables are the same for both analyses, in both case the Gender variable accounts for all the variability in the means. However, the slope coefficients are not the same. The coefficient of EFFECT in the first analysis, -3.188, indicates that the mean for males is 3.188 units less than the average of the male ans female means, whereas the coefficient of DUMMY in the second analysis, -6.375, indicates that the mean of the male scores is 6.375 units less than the mean of the female scores.

**21.2** If we regress on the "nonsense" variable that has levels of 33 and -17, we again get the same ANOVA table, although the value of the slope coefficient changes to -.128. Again, the regression accounts for all the variability in the two group means. If we were only determining whether the effect of gender was significant, we could regress on any of the three types of coding – effect, dummy, or nonsense.

### 21.3

(a) With three levels of the factor, we would need two dummy variables.

(b) The coding is as follows:

| | Coding | | | |
|---|---|---|---|---|
| | Effect | | Dummy | |
| $Y$ | $E1$ | $E2$ | $D1$ | $D2$ |
| 17 | 1 | 0 | 1 | 0 |
| 33 | 1 | 0 | 1 | 0 |
| 26 | 1 | 0 | 1 | 0 |
| 27 | 1 | 0 | 1 | 0 |
| 21 | 1 | 0 | 1 | 0 |
| 11 | 0 | 1 | 0 | 1 |
| 18 | 0 | 1 | 0 | 1 |
| 14 | 0 | 1 | 0 | 1 |
| 18 | 0 | 1 | 0 | 1 |
| 9 | -1 | -1 | 0 | 0 |
| 12 | -1 | -1 | 0 | 0 |
| 10 | -1 | -1 | 0 | 0 |
| 8 | -1 | -1 | 0 | 0 |
| 14 | -1 | -1 | 0 | 0 |

(c) (i) For effect coding the regression output is

```
Dep Var: Y   N: 14   Multiple R: 0.846   Squared multiple R: 0.716
Adjusted squared multiple R: 0.664    Standard error of estimate: 4.335
```

| Effect | Coefficient | Std Error | Std Coef | Tolerance | t | P(2 Tail) |
|---|---|---|---|---|---|---|
| CONSTANT | 16.883 | 1.165 | 0.000 | . | 14.491 | 0.000 |
| E1 | 7.917 | 1.616 | 0.928 | 0.720 | 4.900 | 0.000 |
| E2 | -1.633 | 1.710 | -0.181 | 0.720 | -0.955 | 0.360 |

Analysis of Variance

| Source | Sum-of-Squares | df | Mean-Square | F-ratio | P |
|---|---|---|---|---|---|
| Regression | 521.250 | 2 | 260.625 | 13.866 | 0.001 |
| Residual | 206.750 | 11 | 18.795 | | |

(ii)
```
Dep Var: Y   N: 14   Multiple R: 0.846   Squared multiple R: 0.716
Adjusted squared multiple R: 0.664    Standard error of estimate: 4.335
```

| Effect | Coefficient | Std Error | Std Coef | Tolerance | t | P(2 Tail) |
|---|---|---|---|---|---|---|
| CONSTANT | 10.600 | 1.939 | 0.000 | . | 5.467 | 0.000 |
| D1 | 14.200 | 2.742 | 0.944 | 0.778 | 5.179 | 0.000 |
| D2 | 4.650 | 2.908 | 0.291 | 0.778 | 1.599 | 0.138 |

```
Analysis of Variance
Source       Sum-of-Squares   df   Mean-Square   F-ratio   P
Regression   521.250          2    260.625       13.866    0.001
Residual     206.750          11   18.795
```

Note that the ANOVA tables are exactly the same for the two regressions.

(d) The interpretation of the coefficients for the regression on the effect coding variables are $b_0 = \dfrac{\overline{Y}_{.1}+\overline{Y}_{.2}+\overline{Y}_{.3}}{3}$; $b_1 = \overline{Y}_{.1}-b_0$; and $b_2 = \overline{Y}_{.2}-b_0$.

For the regression on the dummy coding variables, if the reference group, i.e., the group coded by 0's by both variables is group 3, the coefficients are $b_0 = \overline{Y}_{.3}$; $b_1 = \overline{Y}_{.1}-\overline{Y}_{.3}$; and $b_2 = \overline{Y}_{.2}-\overline{Y}_{.3}$.

**21.4** The interpretation is the same whether or not there are equal numbers of scores in each group.

**21.5**

(a) The ANOVA output (using SYSTAT) is:

```
Dep Var: Y   N: 30   Multiple R: 0.696   Squared multiple R: 0.484

Analysis of Variance
Source   Sum-of-Squares   df   Mean-Square   F-ratio   P
A        227.031          1    227.031       3.566     0.071
B        1019.342         2    509.671       8.004     0.002
A*B      104.836          2    52.418        0.823     0.451
Error    1528.167         24   63.674
```

(b) We would need five dummy variables to code the design: one for $A$, two for $B$, and two for the $A \times B$ interaction. The effect coding is presented below:

| | $A$ | $B$ | | $A \times B$ | |
| --- | --- | --- | --- | --- | --- |
| $Y$ | $X1$ | $X2$ | $X3$ | $X4$ | $X5$ |
| 72 | 1 | 1 | 0 | 1 | 0 |
| 63 | 1 | 1 | 0 | 1 | 0 |
| 57 | 1 | 1 | 0 | 1 | 0 |
| 52 | 1 | 1 | 0 | 1 | 0 |
| 69 | 1 | 1 | 0 | 1 | 0 |
| 75 | 1 | 1 | 0 | 1 | 0 |
| 49 | 1 | 0 | 1 | 0 | 1 |
| 71 | 1 | 0 | 1 | 0 | 1 |
| 63 | 1 | 0 | 1 | 0 | 1 |
| 48 | 1 | 0 | 1 | 0 | 1 |
| 40 | 1 | -1 | -1 | -1 | -1 |
| 49 | 1 | -1 | -1 | -1 | -1 |
| 36 | 1 | -1 | -1 | -1 | -1 |

| 50 | 1 | -1 | -1 | -1 | -1 |
| 54 | 1 | -1 | -1 | -1 | -1 |
| 65 | -1 | 1 | 0 | -1 | 0 |
| 45 | -1 | 1 | 0 | -1 | 0 |
| 52 | -1 | 1 | 0 | -1 | 0 |
| 53 | -1 | 1 | 0 | -1 | 0 |
| 57 | -1 | 1 | 0 | -1 | 0 |
| 56 | -1 | 0 | 1 | 0 | -1 |
| 55 | -1 | 0 | 1 | 0 | -1 |
| 49 | -1 | 0 | 1 | 0 | -1 |
| 52 | -1 | 0 | 1 | 0 | -1 |
| 45 | -1 | 0 | 1 | 0 | -1 |
| 57 | -1 | 0 | 1 | 0 | -1 |
| 41 | -1 | -1 | -1 | 1 | 1 |
| 42 | -1 | -1 | -1 | 1 | 1 |
| 57 | -1 | -1 | -1 | 1 | 1 |
| 39 | -1 | -1 | -1 | 1 | 1 |

(c) The correlation matrix for the dummy variables is:

Pearson correlation matrix

| | $X1$ | $X2$ | $X3$ | $X4$ | $X5$ |
|---|---|---|---|---|---|
| $X1$ | 1.000 | | | | |
| $X2$ | 0.000 | 1.000 | | | |
| $X3$ | -0.126 | 0.460 | 1.000 | | |
| $X4$ | 0.082 | 0.100 | 0.051 | 1.000 | |
| $X5$ | 0.042 | 0.062 | -0.048 | 0.465 | 1.000 |

Yes, the dummy variables are correlated with one another. However, the variables belonging to different effects would not be correlated if the design was orthogonal (had an equal number of scores in each cell).

(d) In this case we would want to test hypotheses about unweighted means,

i.e., $H_{0_A}: \mu_1. = \mu_2.$ where $\mu_1. = \dfrac{\mu_{11} + \mu_{12} + \mu_{13}}{3}$

$H_{0_B}: \mu_{\cdot 1} = \mu_{\cdot 2} = \mu_{\cdot 3}$ where $\mu_{\cdot 1} = \dfrac{\mu_{11} + \mu_{21}}{2}$

and $H_{0_{AB}}: \mu_{jk} - \mu_j. - \mu_{\cdot k} - \mu.. = 0$ for all $j, k$. We could test the first of these hypotheses by finding $SS_{A|B,AB} = (R^2_{Y \cdot A,B,AB} - R^2_{Y \cdot B,AB}) SS_Y$, then use

$$F = \frac{(R^2_{Y \cdot A,B,AB} - R^2_{Y \cdot B,AB}) SS_Y / df_A}{(1 - R^2_{Y \cdot A,B,AB}) SS_Y / df_{\text{error}}}.$$

Here, this is $F_A = \dfrac{(1433.20 - 1206.17) / 1}{1528.17 / 24} = 3.56$, $p = .071$

$$F_B = \frac{(1433.20 - 413.86)/2}{1528.17/24} = 8.00, \quad p = .002$$

and $$F_{AB} = \frac{52.42}{63.67} = 0.82, \quad p = .451.$$

The results are identical to those of the standard ANOVA using Type III *SS*.

(e) If we regress only on $X_1$, we get

$$F = MS_A/MS_{\text{error}} = \frac{229.63}{2731.73/28} = 2.354, \quad p = .136. \text{ This a test of the}$$

hypothesis $\mu_{1*} = \mu_{2*}$ where

$$\mu_{1*} = \frac{n_{11}\mu_{11} + n_{12}\mu_{12} + n_{13}\mu_{13}}{n_{11} + n_{12} + n_{13}} \quad \text{and} \quad \mu_{2*} = \frac{n_{21}\mu_{21} + n_{22}\mu_{22} + n_{23}\mu_{23}}{n_{21} + n_{22} + n_{23}}$$

are weighted means. The same result may be obtained by performing an ANOVA but only including $X_1$ as a factor.

**21.6** The results are identical. There is a significant *P* effect, $F(2,32)$ =4.074, $p = .027$.

**21.7** To test for homogeneity of slope, find

$$F = \frac{(R^2_{Y \cdot X, X_1, X_2, X_3, X_4, X_5} - R^2_{Y \cdot X, X_1, X_2, X_3})/2}{(1 - R^2_{Y \cdot X, X_1, X_2, X_3, X_4, X_5})/30} = \frac{(.7121 - .6882)/2}{(1 - .7121)/30} = 1.246.$$

We can't reject the hypothesis that the population regression slopes are the same.

Printed and bound by CPI Group (UK) Ltd, Croydon, CR0 4YY

17/10/2024

01775694-0019